Gene Therapy
Prospective Technology
Assessment in its Societal Context

Cover figure:
Immunocytochemical detection of recombinant STAT1
transcription factor in human fibrosarcoma cells

Cover illustration and illustrations at the beginning of the sections
By Prof. Dr. mult. Thomas Meyer.

Gene Therapy Prospective Technology Assessment in its Societal Context

EDITORS

JÖRG NIEWÖHNER
C:SL-COLLABORATORY: SOCIAL ANTHROPOLOGY
& LIFESCIENCES
INSTITUTE OF EUROPEAN ETHNOLOGY
HUMBOLDT-UNIVERSITY
BERLIN, GERMANY

CHRISTOF TANNERT
RESEARCH GROUP BIOETHICS
AND SCIENCE COMMUNICATION
MAX-DELBRÜCK-CENTER FOR
MOLECULAR MEDICINE (MDC)
BERLIN, GERMANY

ELSEVIER

Amsterdam · Boston · Heidelberg · London · New York · Oxford
Paris · San Diego · San Francisco · Singapore · Sydney · Tokyo

Elsevier
Radarweg 29, PO Box 211, 1000 AE Amsterdam, The Netherlands
The Boulevard, Langford Lane, Kidlington, Oxford OX5 1GB, UK

First edition 2006

Library of Congress Cataloging-in-Publication Data
A catalog record for this book is available from the Library of Congress

British Library Cataloguing in Publication Data
A catalogue record for this book is available from the British Library

ISBN-13: 978-0-444-52806-3

Transferred to digital print 2008

Printed and bound in Great Britain by CPI Antony Rowe, Eastbourne

Contents

Preface xi
Editorial: Building Interdisciplinarity in Research on Gene Therapy xiii
Contributors xxi

Section I Scientific Aspects 1
Identification of Genes Causing Autosomal Recessive Retinitis Pigmentosa 3
Rashid Mehmood, Muhammad Ramzan, Akhtar Ali, Assad Riaz, Fareeha Zulfiqar and Sheikh Riazuddin
1. Introduction 4
2. Materials and Methods 5
3. Results 6
Discussion 10
References 12

Recombinant Adeno-Associated Viral Vectors for CNS Gene Therapy 17
Corinna Burger
1. Introduction 17
2. Biology of AAV 18
3. AAV Gene transfer into the nervous system 19
4. Safety Considerations 22
5. Conclusion 23
References 24

Controlling Adenoviral Gene Transfer in Heart By Catheter-Based Coronary Perfusion 33
J. Michael O'Donnell and E. Douglas Lewandowski
1. Introduction 34
2. Methods 35
3. Results 40
4. Discussion 41
5. Conclusion 44
Acknowledgements 44
References 44

Biological and Cellular Barriers Limiting the Clinical Application of Nonviral Gene Delivery Systems 47

Régis Cartier and Regina Reszka

1. Introduction 48
2. The gene transfer mechanism 48
3. Specificity of targeting peptides during transfection 50
4. The need of standardization 51
5. The effect of physicochemical properties of DNA complexes on transfection 52
6. Understanding the behavior of DNA complexes inside the cell 53
7. Rational design of heteroplexes with *in vivo* application 54

References 55

Designing Polymer-Based DNA Carriers for Non-Viral Gene Delivery: Have We Reached an Upper Performance Limit? 57

Jean-François Lutz

1. Introduction 57
2. How should an ideal carrier behave for gene delivery? 59
3. Basic molecular structure of polymeric carriers for non-viral gene delivery 62
4. Structural improvement of polymeric vectors for gene delivery 65
5. What could be done for improving the situation? 70
6. Summary and last remarks 72

Acknowledgements 72

References 72

Neurogenetic Imaging 77

Merle Fairhurst

1. Introduction 78
2. Background 78
3. Clinical applications 80
4. Ethical considerations and social implications 84
5. Conclusion 86

References 87

Implications of Fetal Gene Therapy for the Medical Profession 89

Thomas Meyer

1. Ethical objections to fetal gene therapy 90
2. *In utero* gene transfer as a new therapeutical option 91
3. The medical profession in the context of fetal gene therapy 92

References 94

Section II Ethical and Legal Aspects 97

Ethical Issues in Gene Therapy Research, An American Perspective 99

Rebecca S. Feinberg

1. Introduction 99
2. Arguments in favor of gene therapy research 100
3. Arguments against gene therapy research 102
4. General societal concerns regarding gene therapy research 104
5. The american approach to gene therapy regulation 106
   References 107

Do Germline Interventions Justify the Restriction of Reproductive Autonomy? 109

Johannes Huber

1. Introduction 110
2. What kind of germline interventions should be investigated and what are some of the related risks? 111
3. A definition of "reproductive autonomy" and possible degrees of its limitation 113
4. Discussion 116
5. Summary 119
   References 120

"Ghost of Christmas Past" ...Eugenics and other Moral Dilemmas Surrounding Genetic Interventions: A Discussion in the Context of Virtue Ethics 123

Agomoni Ganguli

1. Introduction 124
2. The ghost of christmas past 125
3. The ghost of christmas present 126
4. Introducing virtue ethics 129
5. Virtue ethics, therapy and enhancement 132
6. Conclusions 140
   References 141

Biomedical Research and Ethical Regulations in China: Some Observations about Gene Therapy, Human Research, and Struggles of Interest 143

Ole Döring

1. Introduction 144
2. A case of gene therapy – Facts, figures and the emergence of policy 144
3. On cultural expectations 150
4. The role of researchers in ethics regulations and the international state of the art: China's position 153
5. What it means for us? 156
   References 157

Does Patent Granting Hinder the Development of Gene Therapy Products? 159

Clara Sattler de Sousa e Brito

1. Public discussion 160
2. Basic facts about patents 160
3. Patent law and gene therapy 161
4. The concerns 166
5. Patenting conditions and opposition procedures in the case of the BRCA genes 171
6. Conclusion and outlook 174

References and Notes 176

On the Political Side of Gene Therapy, What can be Drawn from the French Situation 181

Anne-Sophie Paquez

1. Gene therapy in the French societal context 182
2. The political stakes of gene therapy: Towards a new French policy-style 187
3. Broadening the scope: Gene therapy from a prospective political point of view 192
4. Conclusion 196

References and Notes 196
Further reading 198
Extracts from Interviews 200

Section III Perception and Communication 201

How to Communicate Risks in Gene Therapy? 203

Andrea T. Thalmann

1. Introduction 204
2. Definitions – clarity with regard to risk, hazards and uncertainty 205
3. Goal of risk communication 206
4. Core problems of risk communication 207
5. Conclusion for communication regarding gene therapy 216

References 217

European Analysis of the Various Procedures Existing to Interrupt a Clinical Research Protocol Thanks to a French Example of Gene Therapy 223

Jacques-Aurélien Sergent, Grégoire Moutel, Josué Feingold, Hervé de Milleville, Eric Racine, Hubert Doucet and Christian Hervé

1. Introduction 224
2. Material and methods 225
3. Results 225
4. Discussion 227
5. Conclusion 227

References 228

The Future of Public Perception of Gene Therapy in Europe, an Educated Guess 229

Markus Schmidt

1. Introduction: Risk perception 230
2. Risk perception towards red and green biotechnology 231
3. Possible factors that will shape future (risk) perception 236
4. Worldviews, fractal discourse and terra incognita 243
5. Brief ideas on management 245

Annex A. Parts of the danish consensus conference on gene therapy (1995), statements on risk, uncertainty and attitudes towards gene therapy 246

References 247

Index 249

Preface

Ladies and Gentlemen,
The Max Delbrueck Center for Molecular Medicine (MDC), Berlin Buch has always performed excellent biomedical research by integrating internationally recognized basic science with the careful development of clinical research and application. Gene therapy exemplifies a field of research and practice that benefits from and even requires this integrative bench-to-bedside approach to unlock its tremendous therapeutic potential. Moreover, science today cannot afford to proceed without engaging in a process of communication and reflection. Transparency in potentially controversial research areas such as gene therapy has always had a high priority at the MDC. Its research group Bioethics and Science Communication, the Life Science Learning Laboratory and the Max Delbrueck Communications Center (MDC.C) vividly attest to the daily interworking of cutting-edge biomedical research, its assessment in a broader social context and its communication to stakeholders and the public.

Thus, it is not unusual for the MDC to host the interdisciplinary workshop, the results of which are presented in this book. I am extremely pleased that the workshop provides an opportunity for young researchers at the beginning of their careers to come together, since they will be able to bring their understanding and appreciation of interdisciplinary questions to future research groups. In my capacity as Chief Executive Officer of the Charité – Universitätsmedizin Berlin, I take great pleasure in welcoming you and lending my support to these kinds of endeavors because they help excellent scientists to develop skills that are of increasing clinical relevance.

I wish the participants of the workshop all the best in their future careers and hope that this volume makes informative and enjoyable reading.

Prof. Detlev Ganten

Charité – Universitätsmedizin Berlin

Editorial

Building Interdisciplinarity in Research on Gene Therapy

Dr. Jörg Niewöhner[1] and Dr. Christof Tannert[2]

[1] C:SL-Collaboratory 'Social Anthropology and the Lifesciences'
Humboldt-University Berlin, Department of European Ethnology
[2] Max Delbrück Center for Molecular Medicine (MDC) Berlin-Buch
Head of the Unit Bioethics and Science Communication

Contents

Acknowledgements xvii
Appendix. Consensus paper xviii
References xix

Perceived by many as one of the most promising areas within the emerging practice of molecular medicine, gene therapy combines expertise from genetics, molecular biology, clinical medicine and human genomics. The Human Genome Project broadly defines gene therapy as "a technique for correcting defective genes responsible for disease development".[1] These techniques include primarily the insertion but also the substitution, repair or regulation of genetic material.

Gene therapy relies and draws heavily on our rapidly expanding knowledge of the role of gene–gene and gene–environment interactions in the pathogenesis of many diseases [1,2]. Newly emerging disciplines such as bioinformatics, and innovative technologies such as functional magnetic resonance imaging (FMRI) (see Fairhurst, this volume) are beginning to provide invaluable information on gene expression *in vivo*. This helps to identify inherited or acquired genetic deficiencies and to reveal their (dys)functionality in disease, thereby isolating potential targets for intervention. Linking these knowledges with other findings in molecular biology,

[1]HGP: http://www.ornl.gov/sci/techresources/Human_Genome/medicine/genetherapy.shtml accessed November 18th, 2005.

biochemistry and material science continually improves our understanding of cellular genomic and metabolic processes. These developments may one day enable the causal treatment of many diseases at the genetic level.

Open any textbook on molecular medicine and gene therapy and you will find a long list of diseases which, the authors promise, will be treated directly or indirectly by means of gene therapy, e.g. cardiovascular diseases, liver diseases, neurological and haematological disorders, cancer, HIV infection and certain forms of arthritis. These are all major diseases that inflict serious pain on patients, often correlate with a range of co-morbidities and in many cases significantly reduce life expectancy or disability-adjusted life years, respectively, while also presenting a substantial financial burden on national health budgets especially in the industrialised world.

Being able to treat these diseases at their roots, in many ways would introduce a new era in medical practice. Rather than focusing on dealing with downstream symptoms, often using rather unspecific drugs, gene therapy may enable clinicians to operate far upstream tackling the genetic defect(s) responsible for the adverse alterations in physiology – in many cases even before they arise. The possibility of upstream interventions will likely foster the development of a form of genetic preventive medicine targeted at the individual or small populations that are sufficiently homogeneous with respect to a particular genetic make-up.

Many members of the research community involved in the development of gene therapy share these visions as far as somatic gene therapy, i.e. the manipulation of genetic material in somatic rather than germline cells, is concerned. Germline interventions intended to produce genetic alterations to be passed on to the next generation present a different topic altogether (Feinberg, this volume). Note that none of the contributors to this volume had any interest in developing germline interventions, nor did they see a need for anyone else to develop that approach to date (Huber, this volume). However, gene therapy *in utero*, i.e. fetal gene therapy "as a third therapeutical option" has been tackled (Meyer, this volume).

Beyond the visions of causal treatment of disease, somatic gene therapy, whether conducted *in* or *ex situ*, faces two major challenges that are being addressed in most current research efforts: gene transfer and appropriate gene expression. Gene transfer ("transfection"), i.e. the delivery of genetic material to the relevant cells, currently relies primarily on the so-called vectors ("gene taxis") and as such is pursued along two different routes: the delivery via the so-called (1) viral vectors (see Burger, this volume, for a review on recombinant adeno-associated viral vectors and O'Donnell and Lewandowski, this volume, on adenoviral gene transfer in the heart) and (2) non-viral or artificial vectors (see Lutz and Cartiers, this volume). All vectors carry their own advantages and disadvantages related to safety and toxicity, DNA carrying capacity, transfection efficiency, availability and costs. Therapeutic gene expression becomes relevant once genetic material has been successfully inserted into the patient's genome and refers to the "frequency of use" of this material. The difficulty lies in achieving a level of expression that leads to an amount of gene product in the body sufficient to abolish the dysfunction without causing significant adverse side effects.

The considerable lack of knowledge about transport, transfection and expression of genes as well as issues related to their long-term functioning means that gene therapy today still carries major risks. Viral material from vectors "going astray" in the body, non-target cells being transfected and significant over- or under-expression of genetic material as well as immunological challenges in the course of the transfection process are difficult to control, yet carry the potential to seriously disrupt physiological processes. The issue of unknowable unknowns aside [3], even a comprehensive and reliable risk assessment remains difficult due to the incomplete understanding of the complexities of gene expression *in vivo*.

In the context of these significant uncertainties, two early applications of gene therapy have caught the public's and the media's attention in Europe as well as the in US. In 1999, a 19-year-old student died during a gene therapy trial in the US due to multiple-organ collapse caused by an immune response against the adenoviral vector because of poorly conceived protocols and malpractice (Feinberg, this volume).[2] In 1999 and 2002 in France (Hôpital Necker, Paris), the gene therapeutic treatment of 11 patients with severe chronic immunodeficiency X1 (SCID-X1) led to their cure but caused leukaemia in three of them (Feinberg, this volume).[3] To the public, these trials demonstrated, first and foremost, gene therapy's serious potential for adverse outcomes. To much of the scientific community, they showed that gene therapy worked in principle, yet at the same time brought home the current lack of comprehensive knowledge. The resulting media coverage and public anxiety and mistrust led to a significant decrease in the enthusiasm related to the development of gene therapy. Despite intensive discussions about possibilities to interrupt a clinical research protocol for clinical, scientific and/or ethical reasons (Sergent, this volume) and more generally about ways of assessing risks associated with gene therapy and re-building critical public trust (Thalmann and Fairhurst, this volume), much of public and private funding for basic research dried up, significantly slowing progress in this area. Promises of gene therapy have persisted and an increasing number of products has entered clinical trials. In China, the first drug based on gene therapy, "Gendicide", has apparently been brought to the market recently (Döring, this volume). Nevertheless, most members of the relevant research communities concede that "... current gene therapy is experimental and has

[2] See also geoscience-online from http://www.g-o.de accessed 16 November 2005.
[3] Also press release Paul Ehrlich Institute from 28 January 2005: http://www.pei.de

not proven very successful in clinical trials. Little progress has been made since the first gene therapy clinical trial ...".[4]

The role of public perception and trust in technology and risk assessment is being emphasised in Europe as well as in the US for some time now [4–7]. Many procedures have been developed to better understand the potential developments of a particular technology and assess its benefits and risks while taking into account public concern. Scenario workshops, Delphi studies and other forms of participatory technology assessment have been described in detail elsewhere (e.g. Ref. [8]) and currently operate in Europe as well as in the US with some success (Schmidt, this volume). Therapy and drug development has so far invoked little public controversy as it has been seen as yielding high benefits while operating within a tightly controlled regulatory regime. While this used to hold for much of medical technology and practice, recent scandals in Europe over blood transfusion, vaccines or organ donation have begun to taint the medical profession's image [9]. Gene therapy, as a practice on the borderline between research and medical practice, presents a somewhat different and more complex case raising a host of ethical issues, from screening and informed consent to negative eugenics or even enhancement (Ganguli and Feinberg, this volume) as well as questions related to public perceptions of risk and trust in regulatory regimes (Schmidt, this volume).

Many members of the natural science as well as the technology assessment community subscribe to the view that these issues should be resolved via ethical reflection and improvement of the public's understanding of science and technology [10]. Some would go further to include members of the public in technology assessment panels in an attempt to broaden the knowledge base with which developments are assessed [11,12]. While these approaches are perceived by most as useful, they implicitly perpetuate the view that scientific practice is able to control and evaluate itself while its outputs need to be debated in a broader context. A subtle critique of this perception of science as removed from and untainted by social practice argues for transparency of scientific process and an emancipation of different kinds of knowledges (e.g. Refs. [13–15]).

This volume is based on an interdisciplinary workshop, which tried to take this critique seriously. Rather than understanding science and medicine as monolithic bodies of knowledge and practice, the workshop was based on an understanding of gene therapy as an epistemic culture [16], i.e. a web of different practices that contribute to the production of a contingent body of knowledge. Accepting this historical and social contingency of present knowledge and practice [17] shifts the emphasis from a

[4] HGP: http://www.ornl.gov/sci/techresources/Human_Genome/medicine/genetherapy.shtml accessed 18 November 2005.

reflection of scientific output by experts external to the particular area of science under consideration, to a focus on scientific practice, which necessarily involves the practitioners themselves. The aim is a change in culture, which introduces into scientific practice a process of reflection enabling the practitioners to appreciate the contingency of their own gaze. It is important to note that science in this case does not only refer to the natural sciences and medicine but includes the social sciences and humanities. The process of understanding the contingency of the own gaze via learning to think differently applies to all those involved in highly specialised disciplines.

This concept in mind, the workshop brought together young post-docs from a wide range of different disciplines for a week explaining and debating their own work as well as interviewing more established researchers in the field. Of course, this can only be a small step. Yet the sessions illustrated that trying to understand each other's work in practice, i.e. excitement, daily routines, constraints, visions and anxieties, can help to de- and re-contextualise one's own work. A *trans*disciplinary group process, that transfers knowledges from one discipline to another, begins to produce *inter*disciplinary individuals able to ask research questions located *in between* disciplines. The book chapters are based on the initial contributions of all participants and reflect an intensive process of internal review and rewriting on the basis of the discussions. They are meant to present the outcome of an interdisciplinary experiment rather than reflecting the entire field of gene therapy. Many issues could not be dealt with. Readers interested in comprehensive reviews of gene therapy are pointed to recent articles [18,19] as well as specific journals in the field.

ACKNOWLEDGEMENTS

This book is the outcome of a European interdisciplinary workshop of young post-doc scientists from the natural and social sciences as well as the humanities, which took place at the Max-Delbrueck-Center for Molecular Medicine (MDC), Berlin Buch, in May 2005. We thank all participants for working extremely hard to put this volume together. Particular thanks goes to Ali ben Salem, whose hard work during and around the workshop has been absolutely essential for realizing this project. The individual chapters represent the participants' written contributions, which were substantially revised on the basis of the discussions. The statement on forthcoming challenges in research and policy was supported by all participants as a consensus paper. Four authors from the US and Pakistan were invited to broaden the European perspective of this book (Mehmood *et al.*; Burger; Feinberg; O'Donnell and Lewandowski, this volume). The workshop was funded by the German Ministry of Education and Research, grant no. BMBF 01 GP 0482.

APPENDIX. CONSENSUS PAPER

We still have a poor understanding of the underlying mechanisms of gene transfer. Public funding is necessary to support the fundamentals of gene therapy.

Gene therapy is now understood as an interdisciplinary research field. Tools have to be developed or adapted from other research areas with an interdisciplinary approach in order to render gene therapy research more efficient (e.g. standardisation, harmonisation, communication).

There is a gap between the availability of molecular diagnostics and that of molecular therapeutics for a variety of diseases.

Consideration should be given to setting limitations to restrict genetic intervention to therapeutic application.

Communication needs to be established and facilitated not only between experts and public but also between the experts themselves, both within the same and between different fields.

Communication of issues related to gene therapy has to respect a balance between transparency and privacy of information, clarity and consistency.

One of the characteristics of gene therapy is that its clinical effects are not fully predictable. It contains various forms of uncertainty and these must be acknowledged and communicated.

Terminology and vocabulary describing gene therapy must be carefully created and used. The concept behind the terminology is of primary importance and should be considered when crossing cultural or national borders. For example, even the term "gene therapy" may create false expectations or fears.

Information should not only include technical aspects about gene therapy but also all processes pertaining to its application (e.g. regulation, conditions, legislation).

Two-way communication is essential as well as constant appreciation of public awareness. The channels through which this bidirectional information is distributed have to be carefully created and constantly refined.

Information should be put forward in an accessible manner, specific to the target audience.

Taking these points into consideration may allow for a better grounding for a truly informed decision base.

Taboos should be questioned in the public debate and legislation process.

There is a need to set a flexible framework to regulate gene therapy in order to adapt to the rapidly changing scientific knowledge and social perception. It is possible to establish an independent regulatory body, which would be able to work on a case-by-case basis. It is very important that this body is not only made up of an expert panel but allows for public hearing and public participation.

The legislative process for gene therapy can be made more flexible and expedient. For example, ensure that laws are revised at regular intervals.

It is important to reach a consensus about terms as well as concepts at the European level while allowing for applications and enforcement to be regulated within the national context.

Review the grounding behind legislation. Do our laws protect what we want them to? What is our concept of life, privacy, risk, appropriate use, individual freedom, and future generation choice? One of the key distinctions is that between therapy and enhancement. Public discussion should be encouraged in order to contribute to the definition of concepts (e.g. through surveys and online consultations).

REFERENCES

[1] E. Jablonka and M. J. Lamb, The changing concept of epigenetics, *Ann, NY Acad Sci.*, 2002, **1**, 82–96.

[2] R. Jaenisch and A. Bird, Epigenetic regulation of gene expression: How the genome integrates intrinsic and environmental signals, *Nat. Genet.*, 2003, **1**, 245–254.

[3] A. Stirling and S. Mayer, *Rethinking Risk*, University of Sussex, SPRU, Brighton, 2000.

[4] R. E. Kasperson, O. Renn, P. Slovic, H. S. Brown, J. Emel, R. Goble, J. X. Kasperson and S. Ratick, The social amplification of risk – A conceptual framework of risk, *Risk Anal.*, 1988, **8**(2), 177–187.

[5] O. Renn, W. J. Burns, J. X. Kasperson, R. E. Kasperson and P. Slovic, The social amplification of risk – Theoretical foundations and empirical applications, *Risk Anal.*, 1992, **48**(4), 137–160.

[6] S. Funtowicz and J. R. Ravetz Post-normal science: An insight now maturing, *Futures*, 1999, **1**, 641–646.

[7] W. Poortingaand N. F. Pidgeon, Exploring the dimensionality of trust in risk regulation, *Risk Anal.*, 2003, **23**(5), 961.

[8] S. Joss and S. Bellucci (eds.), *Participatory Technology Assessment – European Perspectives*, CSD, London, 2002.

[9] M. Fitzpatrick, MMR: Risk, choice, chance, *Br. Med. Bull.*, 2004, **69**, 143–153.

[10] Stifterverband, *PUSH – Public Understanding of Science and Humanities*, Stifterverband für die deutsche Wissenschaft, Bonn, 2000.

[11] S. Joss, Danish consensus conferences as a model of participatory technology assessment: An impact study of consensus conferences on Danish Parliament and Danish public debate, *Sci. Public Policy*, 1998, **25**(1), 2–22.

[12] J. Durant, Participatory technology assessment and the democratic model of the public understanding of science, *Sci. Public Policy*, 1999, **26**(5), 313–319.

[13] A. Irwin and B. Wynne, *Misunderstanding Science? The Public Reconstruction of Science and Technology*, Cambridge University Press, Cambridge, 1996.

[14] B. Wynne, Misunderstood misunderstandings: Social identities and public uptake of science. *Misunderstanding Science* (eds. A. Irwin and B. Wynne), Cambridge University Press, Cambridge, 1996

[15] A. Irwin, Science and citizenship, in *Context and Channels* (eds. E. Scanlon, E. Whitelegg and S. Yates), Routledge, London, 1999, p. 1.

[16] K. Knorr-Cetina, *Epistemic Cultures*, Harvard University Press, Cambridge, MA, 1999.

[17] M. Foucault, *The Order of Things*, Random House, New York, 1970.

[18] F. Anderson, Gene therapy – The best of times, the worst of times, *Science*, 2000, **288**(5466), 627–629.

[19] F. Anderson, Gene therapy scores against cancer, *Nat. Med.*, 2000, **6**(8), 862–863.

Contributors

Numbers in parentheses indicate the pages where the authors' contributions can be found.

Akhtar Ali (3), National Centre of Excellence in Molecular Biology, University of Punjab, Lahore, Pakistan

Clara Sattler de Sousa e Brito (159), Max Planck Institute for Intellectual Property, Competition and Tax Law, Germany

Corinna Burger (17), Department of Molecular Genetics and Microbiology, Powell Gene Therapy Center and McKnight Brain Institute, University of Florida College of Medicine, Gainesville, FL 32610, USA

Régis Cartier (47), Nanotechnology Group, Clinic of Anaesthesiology and Operative Medicine Charite, Berlin University Medical School, Berlin, Germany

J. Michael O'Donnell (33), Program in Integrative Cardiac Metabolism, Center for Cardiovascular Research, University of Illinois at Chicago, College of Medicine, Chicago, IL 60612, USA

Hubert Doucet (223), Bioethics Program, 4333 Queen Mary Street, Montréal University QC Canada

Ole Döring (143), Ruhr University Bochum Faculty for East Asian Studies China's History and Philosophy Building GB 1/137 D-44780 Bochum, Germany

Merle Fairhurst (77), Pain Imaging Neuroscience Group, University of Oxford, Department of Human Anatomy and Genetics, South Parks Road, Oxford OX1 3QX, UK

Rebecca S. Feinberg (99)

Josué Feingold (223), Forensic Medicine and Medical Ethics Lab, 45 Rue des Saints Pères–Medicine Faculty–University René Descartes-Paris 5, France; Honored Director of INSERM Unity 393: "Handicaps génétiques de l'enfant", Hôpital Necker-Enfants Malades Université Denis Diderot – Paris VII France

Agomoni Ganguli (123), Ethics Centre, University of Zurich, Switzerland

Christian Hervé (223), Forsenic Medicine and Medical Ethics Lab, 45 Rue des Saints Pères–Medicine Faculty–University René Descartes-Paris 5, France; International Institut of Research in Ethics and Biomedicine France-Québec, 45 Rue des Saints Pères, Paris, France

Johannes Huber (109), Faculty of Medicine, Ludwig-Maximilians-University, Munich and Institute for Scientific Issues Related to Philosophy and Theology, School of Philosophy S.J., Munich, Germany

E. Douglas Lewandowski (33), Program in Integrative Cardiac Metabolism, Center for Cardiovascular Research, University of Illinois at Chicago, College of Medicine, Chicago, IL 60612, USA

Jean-François Lutz (57), Research Group Nanotechnology for Life Science, Fraunhofer Institute for Applied Polymer Research, Geiselbergstrasse 69, Golm 14476, Germany

Rashid Mehmood (3), National Centre of Excellence in Molecular Biology, University of Punjab, Lahore, Pakistan

Thomas Meyer (89), Leibniz-Forschungsinstitut für Molekulare Pharmakologie, Berlin-Buch, Germany

Hervé de Milleville (223), Laboratoire en Sciences de Systèmes d'Information (LASSI), Director, EISTI, Cergy-Pontoise, France

G. Moutel (223), Forensic Medicine and Medical Ethics Lab, 45 Rue des Saints Pères–Medicine Faculty–University René Descartes-Paris 5, France; International Institut of Research in Ethics and Biomedicine France-Québec, 45 Rue des Saints Pères, Paris, France

Anne-Sophie Paquez (181), Sociology and Public Policy Program, Institut d'Etudes Politiques (Sciences Po), Paris

Eric Racine (223), Bioethics Program, 4333 Queen Mary Street, Montréal University QC Canada

Muhammad Ramzan (3), National Centre of Excellence in Molecular Biology, University of Punjab, Lahore, Pakistan

Regina Reszka (47), Helios Research Center, Wiltbergstrasse 50, 13125 Berlin-Buch, Germany

Assad Riaz (3), National Centre of Excellence in Molecular Biology, University of Punjab, Lahore, Pakistan

Sheikh Riazuddin (3), National Centre of Excellence in Molecular Biology, University of Punjab, Lahore, Pakistan

Markus Schmidt (229), Institute of Risk Research, University of Vienna, Austria

Jacques-Aurélien Sergent (223), Groupe de Recherche sur les Physiopathologies Hépatiques Héréditaires (GRP2H), Univesité de Cergy-Pontoise, France; Forensic Medicine and Medical Ethics Lab, 45 Rue des Saints Pères–Medicine Faculty–University René Descartes-Paris 5, France; International Institut of Research in Ethics and Biomedicine France-Québec, 45 Rue des Saints Pères, Paris, France

Andrea T. Thalmann (203), Swiss Federal Office for Public Health, Schwarztorstr. 96, CH-3003 Bern, Switzerland

Fareeha Zulfiqar (3), National Centre of Excellence in Molecular Biology, University of Punjab, Lahore, Pakistan

Protective effects of interferons on virus replication demonstrated in a cytopathic effect assay with virus-infected cells

Section I
Scientific Aspects

Identification of Genes Causing Autosomal Recessive Retinitis Pigmentosa

Rashid Mehmood, Muhammad Ramzan, Akhtar Ali, Assad Riaz, Fareeha Zulfiqar and Sheikh Riazuddin

National Centre of Excellence in Molecular Biology, University of Punjab, Lahore, Pakistan

Abstract

Retinitis pigmentosa (RP) is a group of inherited disorders in which abnormalities of the photoreceptors or the retinal epithelium (RPE) of the retina lead to progressive visual loss. RP can be inherited in autosomal dominant, autosomal recessive, X-linked, mitochondrial and genetically more complex modes. The purpose of the study was to identify genes causing autosomal recessive retinitis pigmentosa. In this study, 50 families showing autosomal recessive mode of inheritance were screened for their linkage to the already reported candidate genes/loci by amplifying the STS markers flanking the genes, followed by their genotyping and haplotype analysis. Two of the families PKRP021 and PKRP019 showed linkage to PDE6A gene, which encodes the subunit of cGMP phosphodiesterase, the effector of the phototransduction cascade. One family, PKRP 014 was found linked with CNGB1 gene, whose product is a subunit of the cGMP-gated cation channel. Fundus examination and patients' history revealed typical symptoms of progressive photoreceptor degeneration. The study identified the carrier status of normal individuals of the families, which is of great help in genetic counseling. The study may be useful for presymptomatic and prenatal diagnosis of this genetic disease. To do this, however, enrollment of new families with RP is required thereby allowing for the identification of causative genes of RP so as to offer preventive measures and novel therapies to affected individuals as well as to understand vision process in a better way.

Keywords: retinal diseases, retinitis pigmentosa, linkage analysis, genetic diseases, gene therapy.

Contents

1. Introduction 4
2. Materials and Methods 5
   2.1 Enrollment of families and clinical evaluation 5
   2.2 Blood collection and DNA extraction 5
   2.3 PCR for microsatellites 6
   2.4 Preparation of samples for ABI 3100 genetic analyzer 6
   2.5 Haplotype analysis and LOD score calculation 6
3. Results 6
   3.1 PKRP021 7
   3.2 Clinical evaluation 7
   3.3 PKRP019 8
   3.4 PKRP014 9
Discussion 10
References 12

1. INTRODUCTION

Retinal photoreceptor dystrophies are a clinically and genetically heterogeneous group of retinal degenerations that together form the most frequent cause of inherited visual disorders, with an estimated prevalence of 1 in 4000 [1], and are responsible for the visual handicaps of 1.5 million individuals worldwide [2–5]. In RP, photoreceptor cells (rods and cones in the neurosensory retina) degenerate, leading to the loss of photoreceptor function. The clinical features of RP include night blindness, constriction and gradual loss of peripheral visual field followed by eventual loss of central vision, and the clinically visible abnormal pigmentation that frequently accompanies the death of photoreceptors, creating a bone spicule-like appearance [6]. Other features include narrowed retinal vessels, depigmentation of the retinal pigment epithelium, waxy pallor of the optic discs, vitreous cells, and the loss of photoreceptors or photoreceptors with shortened or absent outer segments on histopathologic study [6]. Patients with advanced forms of disease can be identified by very small or nondetectable electroretinograms (ERGs). Posterior subcapsular cataracts, refractive errors such as astigmatism and myopia, and some cystoid macular edema develop in many cases of RP. The course of RP is intractable, and there is no effective treatment at present.

Gene mapping and gene discovery have revealed that the molecular genetic causes of RP are unusually complicated [7]. Genes associated with RP encode proteins that are involved in phototransduction, the visual cycle (production and recycling of the chromophore of rhodopsin), retinal metabolism and cell–cell interaction, intracellular transport proteins, splicing factors, photoreceptor structure, and photoreceptor cell transcription factors [8,9]. Most retinal disease genes are preferentially expressed in photoreceptors. In addition, some of the genes linked to retinal diseases are essential for normal retinal development [10]. However, the function of many genes associated with RP remains unknown.

RP is clinically and genetically heterogeneous with a wide variation in severity, mode of inheritance, age of onset, progression, and phenotype. Most cases are not associated with other extraocular abnormalities (nonsyndromic RP). In some families, RP is a manifestation of other diseases such as Usher syndrome, Kearns-Sayre syndrome, and the Bardet–Biedl syndrome [5]. It can be inherited in autosomal dominant, autosomal recessive, X-linked, mitochondrial, or as a digenic trait [8]. It has been estimated that autosomal dominant RP (adRP) and autosomal recessive RP (arRP) each accounts for approximately 20% of the RP cases, and that X-linked RP (xlRP) accounts for 10% of the cases [11,12]. Digenic RP is rare. Approximately 50% of all RP cases are defined as simplex RP, which includes sporadic RP and a great proportion of arRP [12,13]. With the application of molecular genetic technologies to RP in the early 1980s, at least 40 genes responsible for nonsyndromic RP have been mapped or identified, and many more remain to be found [8,14–16].

Although isolation of causative genes for RP has comparatively progressed in the past few years following the sequencing of the human genome, the molecular genetics of RP still needs elaboration. The extreme genetic heterogeneity seriously hinders linkage analysis, so studying large consanguineous families afflicted with RP with more affected individuals has proved helpful. As a consequence, families with multiple affected individuals showing clear segregation will be effective for the linkage studies. In such studies, Pakistan can play an important role because of traditional consanguineous marriages within the same ethnic groups [17]. These families provide excellent resource material for conventional linkage analysis and refining candidate interval for positional cloning of gene causing RP.

With this objective in mind, the present study was embarked upon to carry out linkage analysis of RP causing genes in large consanguineous Pakistani families. Genetic testing will enable presymptomatic and prenatal diagnosis of the members of such families. Moreover, carrier diagnosis can be offered for individuals whose family members are known to be associated with a particular locus. It will permit a more accurate genetic counseling to families. This research will ultimately pave the way for early diagnosis of disease and the development of new specific therapies.

2. MATERIALS AND METHODS

2.1. Enrollment of families and clinical evaluation

Families with two or more affected individuals were contacted through 'Hospitals & Centers for special education (for Blind)'. A thorough medical history, ophthalmic history, and physical examination focusing on features associated with syndromic RP were obtained to establish the cause of retinitis pigmentosa in a person with RP and rule out the possibility of environmental influence. Clinically, the affected individuals were tested for: (1) visual acuity, (2) color vision using standard Ishihara plates, (3) peripheral fields with a tangent screening method, and (4) retinal examination to include black pigment in the peripheral retina, macular atrophy, significant macular discoloration, and vessels narrowing in nasal/temporal retina.

2.2. Blood collection and DNA extraction

Blood 5–10 ml samples were drawn and collected in 400 μl EDTA (anticoagulant) from all the affected individuals, their normal siblings, parents, and grand parents to trace the mode of inheritance. Genomic DNA was extracted by a nonorganic method [18]. DNA concentration was determined by agarose gel electrophoresis and spectrophotometry. Working DNA concentration was kept at 25 ng/μl and 2 μl of DNA was used in 5 μl PCR volume.

2.3. PCR for microsatellites

PCR was carried out in 96-well microtitre plates. Replicate plates were made by dispensing 2 μl aliquots containing 50 ng of DNA. For PCR amplification of microsatellites, 5 μl reaction volume was used. About 3–6 fluorescently labeled primers (with forward primer labeled with one of the three fluorescent dyes – FAM, VIC, NED) were used for linkage analysis of loci. The markers used for each linkage encompassed the chromosomal locations reported for RP loci (http://genome.ucsc.edu/cgi-bin/hgGateway). The markers were mostly dinucleotide repeats and were chosen from the Marshfield Comprehensive Human Genetic Maps (http://www.marshmed.org/genetics/) for each chromosome. The primer sequences for amplification of each marker are listed in the genome database (http://genome.ucsc.edu/cgi-bin/hgGateway).

2.4. Preparation of samples for ABI 3100 genetic analyzer

For fluorescently labeled primers, with forward primer labeled with one of the three fluorescent dyes – FAM, VIC or NED, LIZ internal size standard (Perkin Elmer) was used to run on 3100 Genetic Analyzer. PCR products of different sizes were pooled (maintaining the difference of 25 nucleotides in case the products are labeled with similar dye) by adding 1.0–1.5 μl of the PCR product in 11.8 μl of deionized formamide containing 0.2 μl of LIZ size standard (Perkin Elmer). The samples were denatured at 95°C for 5 min and quickly chilled by keeping on ice for 2–3 min before running in the ABI prism 3100 genetic analyzer.

2.5. Haplotype analysis and LOD score calculation

Haplotype analysis was carried out to confirm or, otherwise, reject the linkage of the affected family to a particular locus. Two-point LOD scores between arRP and the markers in the linked region were calculated with the aid of facility available at Bioinformatics lab.

3. RESULTS

Fifty families were included in the study. Two families, PKRP021 and PKRP019, showed linkage to PDE6A gene while family PKRP014 was found linked with CNGB1 gene.

3.1. PKRP021

This family was enrolled from district Kasur, Punjab. This family consists of 3 loops. Three affected and four normal individuals were included in the study. Their blood samples were collected and the PCR carried out. The family was typed for the following seven markers. D5S436, D5S812, D5S2090, D5S2013, D5S2015, D5S1469, and D5S410. As is clear from the haplotype in Fig. 1, recombination phenomenon occurred between the markers D5S2090 and D5S2013, thus defining the critical region below 150.34 cM. This is the region where the gene PDE6A actually lies. All the affected patients are showing homozygosity through descent for all markers below D5S2090 (150.34 cM), while the normal individuals appear to be carriers. The genotypes of all the individuals are shown in the pedigree of Fig. 1.

Two-point LOD scores between arRP and the markers in the linked region were calculated. Maximum LOD score 2.55891 was obtained for the markers D5S2090 and D5S2015, a value which is consistent with linkage.

3.2. Clinical evaluation

The patients showing linkage with PDE6A were clinically evaluated. A history of night blindness since early childhood was reported. Visual field testing revealed marked peripheral field loss. Fundus examination (Fig. 2) showed abnormalities, typical of RP, including attenuated retinal vessels and

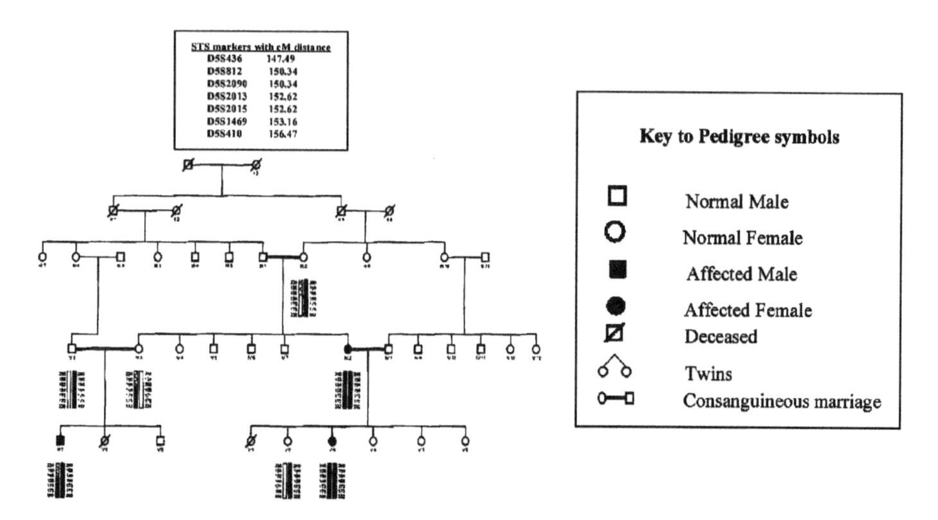

Fig. 1. Pedigree and Haplotype of family PKRP021. Affected individuals are homozygous for mutant allele, while other typed members are carriers of the mutant allele.

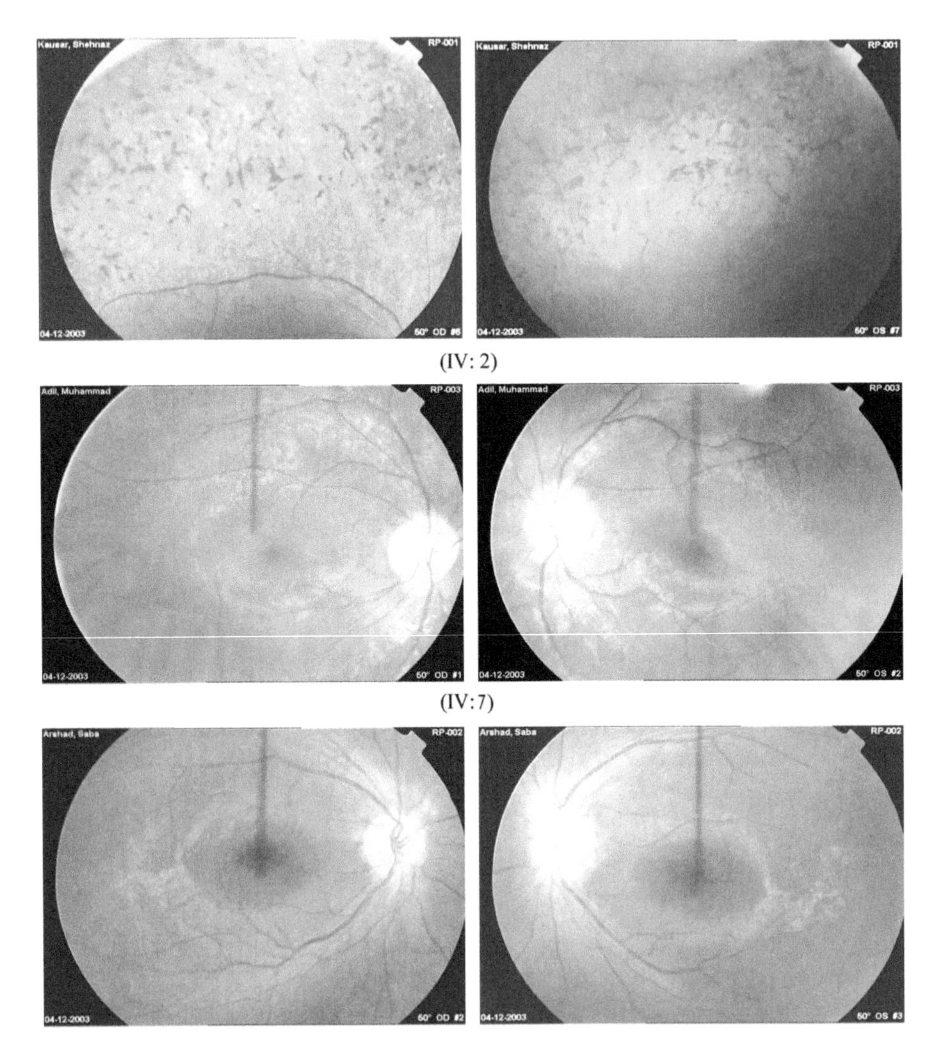

Fig. 2. Retinal photographs of family PKRP021. As RP is mostly progressive in nature, so the older patient (IV:2 in the figure) has extensive pigmentation in retina with complete visual loss. The younger patients (V:7 and V:3) showing initial symptoms of RP in the retina had partial impaired vision.

bone-spicule pigments in the midperipheral retina. Younger patients (V:7 and V:3) revealed symptoms of progressive RP.

3.3. PKRP019

This family was enrolled from district Kasur, Punjab. It was a large inbred family, consisting of three loops. Four affected patients and 10 normal individuals were included in the study. After DNA extraction from their blood, the individuals were typed for the following markers: D5S436, D5S812, D5S2013, D5S2015, and D5S1469. All the affected patients were homozygous by descent. Rest of the individuals appear to be the carriers of the defective gene. Recombination event can be observed between the markers D5S436 and D5S812, in individuals VI:8 and VII:1. All of the affected individuals were homozygous below D5S436, the region where PDE6A is actually located, thus demonstrating the linkage of the family to PDE6A gene. The homozygosity in affected patients and carrier status of the rest of the members is shown in Fig. 3.

3.4. PKRP014

This family was also enrolled from Kasur, Punjab. It was a two-looped consanguineous family with three affected patients. Three affected and nine normal individuals were typed for the following markers: D16S3140, D16S3057, and D16S494. The homozygous status of the affected individuals is evident from the pedigree shown in Fig. 4. The carrier status of parents and other individuals is also apparent. The affected individuals were also clinically evaluated. Patients' history, visual testing, and ophthalmic examination revealed typical symptoms of RP.

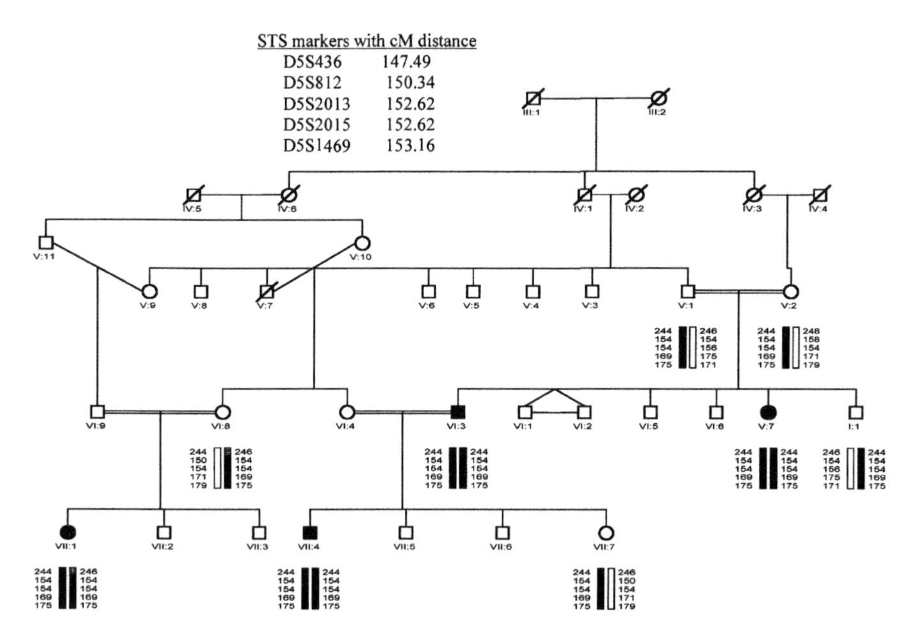

Fig. 3. Pedigree and Haplotype of family PKRP019. The affected individuals are homozygous for mutated allele, while the rest of the typed individuals appear to be the carriers.

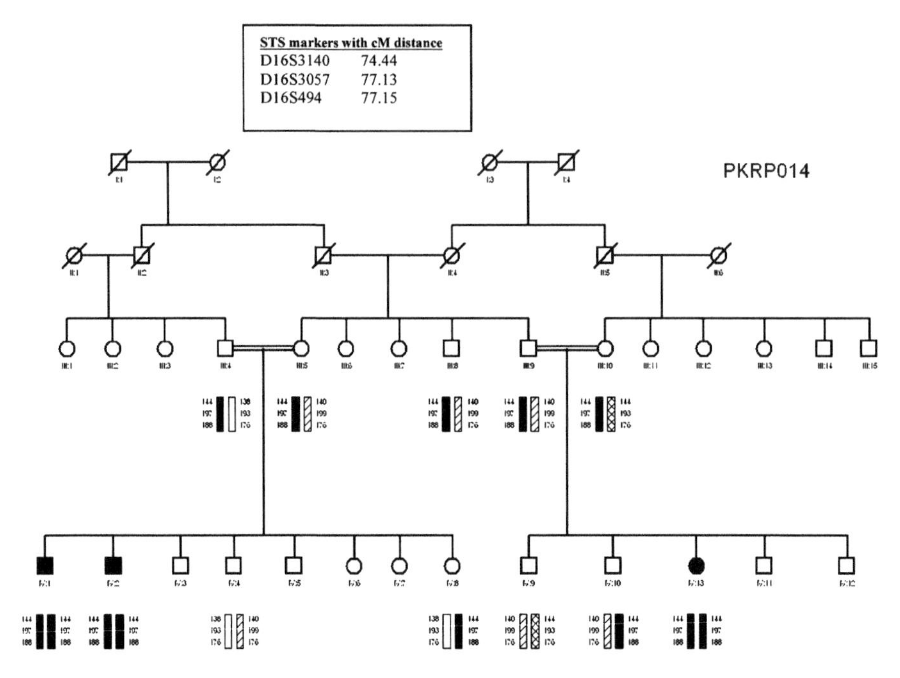

Fig. 4. Pedigree and Haplotype of family PKRP014. Of the typed individuals, all three affected individuals are homozygous for the mutant allele, seven individuals are carriers and two individuals carry both normal alleles.

DISCUSSION

In addition to demonstrating the power of the homozygosity mapping approach in localizing rare autosomal recessive disease loci, the data emerging from studies of a large number of consanguineous arRP families suggest that there is a considerable degree of genetic heterogeneity in the disease [19], and that probably no single gene is responsible for a significant proportion of arRP cases. The extensive genetic heterogeneity found in inherited disorders greatly hinders the finding of loci responsible for these diseases [20].

In summary, the results from this study show that among 50 families with arRP, two families were found to be linked with PDE6A gene and one with CNGB1 gene. Linkage of Pakistani RP families with these loci was not reported in the past. Clinical tests of the family PKRP021 exhibited the typical symptoms of progressive RP. A history of night blindness starting in early childhood was reported. Visual field testing revealed marked peripheral field loss. Fundus examination (as shown in Fig. 2) showed abnormalities typical of RP including attenuated retinal vessels, and bone-spicule pigments in the midperipheral retina. These abnormalities are typical indicators of a progressive photoreceptor degeneration [3,21]. It is worth pointing out that the affected members of both inbred families (PKRP019 and PKRP021) share the same allele at D5S1469 (175/175).

It can be estimated from these results that 4% of cases of recessive RP are caused by mutation in PDE6A. The data are quite close to those described by Dryja *et al.* [22], for the same gene. The estimated prevalence of 4% for PDE6A mutations among families with recessive RP is similar to the 4% prevalence estimated for PDE6B by summing the data from three groups i.e. 4 of 92 families [23], 3 out of 19 families [24], and 2 out of 101 families [25].

Mutations in the PDE6A gene encoding the α subunit of phosphodiesterase cause retinal degeneration in Welsh Cardigan Corgi dogs [26]. The mechanisms by which PDE6A and PDE6B mutations lead to RP are probably similar. Two animal models – the rd mouse and Irish Setter dogs with PDEB mutations have exhibited reduced PDE activity and high levels of cGMP [27,28]; While the elevated levels of cGMP are thought to be deleterious to photoreceptor cells [29,30], the exact mechanism by which mutated PDE6A and PDE6B bring about photoreceptor degeneration needs elaboration. As hypothesized by [22], cGMP concentration gets elevated because of the absent activity of cGMP-phosphodiesterase in the mutant photoreceptor cells. The raised cGMP levels become toxic to photoreceptor cells. Alternatively, the elevated concentration of cGMP may increase the proportion of open cGMP-gated channels in the rod outer segment membrane. The high cGMP levels should result in a higher-than-normal proportion of open channels and a presumably toxic increase in the influx of sodium and calcium ions into the cytoplasm.

The frequency of linkage to CNGB1 gene from this study was found to be 2% (1 out of 50 families). To date, only two mutations have been reported in CNGB1 gene (encoding cGMP-gated cation channel (β-subunit) [31,32], while seven mutations have been reported in CNGA1 gene (encoding cGMP-gated cation channel (α-subunit) [33–35].

The results of the present study and others [36,37] support the approach of screening regions of the genome that contain candidate genes, prior to a systematic genomewide search, mainly because of the cost/benefit ratio of the projects. This would be the case for diseases with genetic heterogeneity, where a great number of nuclear families and a high density of markers (spaced every 5 cM) would be necessary in order to allow us to obtain linkage through genomewide screening [38].

Once the causative gene has been identified, gene therapy carries great potential for treating retinal and choroidal diseases [39]. As many genes responsible for inherited eye disorders within the retina have been identified, diseases of the eye are prime candidates for this form of therapy. The eye, as a well-characterized restricted site, also has the advantage of being highly accessible with altered immunological properties; both important considerations for easy delivery of virus and avoidance of systemic immune responses [40]. Recessive degenerations, which result from lack of function of a gene product, are good candidates for gene replacement therapy. Photoreceptor rescue has been achieved by subretinal transformation of β-PDE. This transformation resulted in delayed photoreceptor degeneration [41,42].

As genetic diseases are particularly common in Pakistan because of traditional consanguineous marriages, the genetic resource in Pakistan can be exploited to conduct research in the area of genetic diseases and genetic diversity. Research is being conducted at various institutes aiming at various aspects of genetics in collaboration with well renowned institutes like NIH, NEI, IOWA, UCL etc. Research has been especially focused on the genetic characterization of deafness, Usher syndrome, Pendred syndrome, retinitis pigmentosa, cataract, glaucoma, baldness, leber congenital amaurosis, asthma, cone-rod dystrophy, congenital microphthalmia, stuttering, thalassaemia, and mental retardation. The extensive research on these genetic diseases led to the identification of novel genes/loci responsible for these disorders. Apart from these genetic diseases, the research has also been expanded to include investigation of Y-chromosomal microsatellites, HLA polymorphism, and the study of the genotypes of different ethnic groups for forensic purposes.

A comprehensive genetic study is a pre-requisite for gene therapy trials. Until now the objective of genetic research has been the genetic counseling and prenatal diagnosis. Gene therapy seems to be far fetched but the great scope of the therapeutic alternative seems to be progressing ever forward. Gene therapy holds great promise to cure genetic diseases (which are tragically prevalent in Pakistan). For this, more research is obviously required to understand the etiology of genetic diseases. Population studies are important to enroll more families in the genetic research so as to identify the causative genes. In conclusion, knowledge of the disease causing genes remains one of the primary aims of genetic research in genetic diseases as it seems likely to be a key element in better understanding the diseases, future diagnostics, and novel therapies (including gene therapy) to combat their incidence.

REFERENCES

[1] M. Jay, Figures and fantasies: The frequencies of the different genetic forms of retinitis pigmentosa, *Birth Defects Orig. Artic. Ser.*, 1982, **18**, 167–173.

[2] M. Haim, N. V. Holm and T. Rosenberg, Prevalence of retinitis pigmentosa and allied disorders in Denmark. I. Main results, *Acta Ophthalmol. (Copenhagen).*, 1992, **70**, 178–186.

[3] E. L. Berson, Retinitis pigmentosa. The Friedenwald Lecture. Invest, *Ophthalmol. Vis. Sci.*, 1993, **34**, 1659–1676.

[4] R. Kumar-Singh, G. J. Farrar, F. Mansergh, P. Kenna, S. Bhattacharya, A. Gal and P. Humphries, Exclusion of the involvement of all known retinitis pigmentosa loci in the disease present in a family of Irish origin provides evidence for a sixth autosomal dominant locus (RP8), *Hum. Mol. Genet.*, 1993, **2**, 875–878.
[5] T. P. Dryja and T. Li, Molecular genetics of retinitis pigmentosa, *Hum. Mol. Genet.*, 1995, **4**, 1739–1743.
[6] J. R. Heckenlively, S. L. Yoser, L. H. Friedman and J. J. Oversier, Clinical findings and common symptoms in retinitis pigmentosa, *Am. J. Ophthalmol.*, 1988, **105**, 504–511.
[7] C. Rivolta, D. Sharon, M. M. DeAngelis and T. P. Dryja, Retinitis pigmentosa and allied diseases: Numerous diseases, genes, and inheritance patterns, *Hum. Mol. Genet.*, 2002, **11**, 1219–1227.
[8] J. K. Phelan and D. Bok, A brief review of retinitis pigmentosa and the identified retinitis pigmentosa genes, *Mol. Vis.*, 2000, **6**, 116–124.
[9] M. M. Hims, S. P. Diager and C. F. Inglehearn, Retinitis pigmentosa: Genes, proteins and prospects, *Dev. Ophthalmol.*, 2003, **37**, 109–125.
[10] J. Lord-Grignon, N. Tetreaultm, A. J. Mearsm, A. Swaroopm and G. Bernierm, Characterization of new transcripts enriched in the mouse retina and identification of candidate retinal disease genes, *Invest. Ophthalmol. Vis. Sci.*, 2004, **45**(9), 3313–3319.
[11] J. A. Boughman, P. M. Conneally and W. E. Nance, Population genetic studies of retinitis pigmentosa, *Am. J. Hum. Genet.*, 1980, **32**, 223–235.
[12] M. Jay, On the heredity of retinitis pigmentosa, *Br. J. Ophthalmol.*, 1992, **66**, 405–416.
[13] M. Ham, Retinitis pigmentosa: Problems associated with genetic classification, *Clin. Genet.*, 1993, **44**, 62–70.
[14] C. F. Inglehearn, Molecular genetics of human retinal dystrophies, *Eye*, 1998, **12**, 571–579.
[15] S. Van Soest, A. Westerveld, P. T. de Jong, E. M. Bleeker-Wagemakers and A. A. Bergen, Retinitis pigmentosa: Defined from a molecular point of view, *Surv. Ophthalmol.*, 1999, **43**, 321–334.
[16] A. Rattner, H. Sun and J. Nathans, Molecular genetics of human retinal disease. *Annu. Rev. Genet.*, 1999, **33**, 89–131.
[17] S. A. Shami and L. H. Schmitt, Consanguinity related prenatal and postnatal mortality of the populations of seven Pakistani Punjab cities, *J. Med. Genet.*, 1989, **26**, 267–271.
[18] J. Grimberg, S. Nawoschik, L. Belluscio, R. McKee, A. Turck and A. Eisenberg, A simple and efficient non-organic procedure for the isolation of genomic DNA from blood, *Nucleic Acid Res.*, 1989, **17**, 8390.
[19] U. Finckh, S. Xu, G. Kumaramanickavel, M. Schurmann, J. K. Mukkadan, S. T. Fernandez, S. John, J. L. Weber, M. J. Denton and A. Gal, Homozygosity mapping of autosomal recessive retinitis pigmentosa locus (RP22) on chromosome 16p12.1–p12.3. *Genomics*, 1998, **48**, 341–345.
[20] T. Strachan and A. P. Read (eds.), *Human Molecular Genetics,* BIOS Scientific, Oxford, 1996.
[21] S. H. Huang, S. J. Pittler, X. Huang, L. Oliveira, E. L. Berson and T. P. Dryja, Autosomal recessive retinitis pigmentosa caused by mutations in the alpha subunit of rod cGMP phosphodiesterase, *Nat. Genet.*, 1995, **11**, 468–471.
[22] T. P. Dryja, E. R. David, H. C. Sherleen and L. B. Eliot, Frequency of mutations in the gene encoding the α subunit of rod cGMP-phosphodiesterase in autosomal recessive retinitis pigmentosa, *Invest. Ophthalmol. Vis. Sci.*, 1999, **40**, 1859–1865.
[23] M. E. McLaughlin, T. L. Ehrhart, E. L. Berson and T.P. Dryja, Mutation spectrum of the gene encoding the β subunit of rod phosphodiesterase among patients with autosomal recessive retinitis pigmentosa, *Proc. Natl. Acad. USA*, 1995, **92**, 3249–3253.

[24] M. Danciger, V. Heilbron, Y. Q. Gao, D. Y. Zhao, S. G. Jacobson and D. B. Farber, A homozygous PDE6B mutation in a family with autosomal recessive retinitis pigmentosa, *Mol. Vis.*, 1996, **2**, 10.
[25] A. Veske, U. Orth and K. Rüther, Mutations in the gene for the β-subunit of rod photoreceptor cGMP-specific phosphodiesterase (PDEB), in patients with retinal dystrophies and dysfunctions (ed. R.E. Anderson), *Degenerative Diseases of the Retina*, Springer, Berlin, 1995, pp. 313–322. Advanced clinical research, for ophthalmologists. Proceedings of the Sixth International Symposium on Retinal Degenerations, November 4–9, 1994, Jerusalem, Israel.
[26] S. M. Petersen-Jones, D. Entz and D. R. Sargan, Mutation linked to cGMP-phosphodiesterase alpha gene causes generalised progressive retinal atrophy in the Cardigan Welsh Corgi, *Invest. Ophthalmol. Vis. Sci.*, 1998, **39** (4), S880.
[27] M. L. Suber, S. J. Pittler and N. Qin, Irish setter dogs affected with rod/cone dysplasia contain a nonsense mutation in the rod cGMP phosphodiesterase β-subunit gene, *Proc. Natl. Acad. Sci. USA*, 1993, **90**, 3968–3972.
[28] D. B. Farber and R. N. Lolley, Enzymic basis for cyclic GMP accumulation in degenerative photoreceptor cells of mouse retina, *J. Cyc. Nucl. Res.*, 1976, **2**, 139–148.
[29] E. N. Lolley, D. B. Farber, M. E. Rayborn and J. G. Hollyfield, Cyclic GMP accumulation causes degeneration of photoreceptor cells: Simulation of an inherited disease. *Science*, 1977, **196**, 664–666.
[30] R. J. Ulshafer, C. A. Garcia and J. G. Hollyfield, Sensitivity of photoreceptors to elevated levels of cGMP in the human retina, *Invest. Ophthalmol. Vis. Sci.*, 1980, **19**, 1236–1241.
[31] C. Bareil, C. P. Hamel, V. Delague, B. Arnaud, J. Demaille and M. Claustres, Segregation of a mutation in CNGB1 encoding the β-subunit of the rod cGMP-gated channel in a family with autosomal recessive retinitis pigmentosa, *Hum. Genet.*, 2001, **108**(4), 328–334.
[32] H. Kondo, M. Qin, A. Mizota, M. Kondo, H. Hayashi, K. Hayashi, K. Oshima, T. Tahira and K. Hayashi, A homozygosity-based search for mutations in patients with autosomal recessive retinitis pigmentosa, using microsatellite markers, *Invest. Ophthalmol. Vis. Sci.*, 2004, **45**(12), 4433–4439.
[33] Q. Zhang, F. Zulfiqar, S. A. Riazuddin, X. Xiao, Z. Ahmad, S. Riazuddin and J. F. Hejtmancik, Autosomal recessive retinitis pigmentosa in a Pakistani family mapped to CNGA1 with identification of a novel mutation, *Mol. Vis.*, 2004, **10**, 884–889.
[34] T. P. Dryja, J. T. Finn, Y. W. Peng, T. L. McGee, E. L. Berson and K. W. Yau, Mutations in the gene encoding the alpha subunit of the rod cGMP-gated channel in autosomal recessive retinitis pigmentosa, *Proc. Natl. Acad. Sci. USA*, 1995, **92**, 10177–10181.
[35] E. Paloma, A. Martinez-Mir, B. Garcia-Sandoval, C. Ayuso, L. Vilageliu, R. Gonzalez-Duarte and S. Balcells, Novel homozygous mutation in the alpha subunit of the rod cGMP-gated channel (CNGA1) in two Spanish sibs affected with autosomal recessive retinitis pigmentosa, *J. Med. Genet.*, 2002, **39**, E66.
[36] K. Fukai, J. Oh, M. A. Karim, K. J. Moore, H. H. Kandil, H. Ito and J. Burger, Homozygosity mapping of the gene for Chediak–Higashi syndrome to chromosome 1q42–q44 in a segment of conserved synteny that includes the mouse beige locus (bg), *Am. J. Hum. Genet.*, 1996, **59**, 620–624.
[37] E. Pastural, F. J. Barrat, R. Dufourcq-Lagelouse, S. Certain, O. Sanal, N. Jabado and R. Seger, Griscelli disease maps to chromosome 15q21 and is associated with mutations in the myosin-Va gene, *Nat. Genet.*, 1997, **16**, 289–292.
[38] M. Gshwend, O. Levran, L. Kruglyak, K. Ranade, P. C. Verlander, S. Shen and S. Faure, A locus for Fanconi anemia on 16q determined by homozygosity mapping, *Am. J. Hum. Genet.*, 1996, **59**, 377–384.

[39] P. A. Campochiaro, Gene therapy for retinal and choroidal diseases, *Biol. Ther.*, 2002, **2**(5), 537–544.
[40] N. S. Dejneka and J. Bennett, Gene therapy and retinitis pigmentosa: Advances and future challenges. *Bioessays*, 1991, **23**(7), 662–668.
[41] M. Takahashi, H. Miyoshi, I. M. Verma and F. H. Gage, Rescue from photoreceptor degeneration in the rd mouse by human immunodefiency virus vector-mediated gene transfer, *J. Virol.*, 1999, **73**, 7812–7816.
[42] R. Kumar-Singh and D. B. Farber, Encapsidated adenovirus mini-chromosome mediated delivery of genes to the retina: Application to the rescue of photoreceptor degeneration, *Hum. Mol. Genet.*, 1998, **7**, 1893–1900.
http://genome.ucsc.edu/cgi-bin/hgGateway, http://www.marshmed.org/genetics/

Recombinant Adeno-Associated Viral Vectors for CNS Gene Therapy

Corinna Burger

Department of Molecular Genetics and Microbiology, Powell Gene Therapy Center and McKnight Brain Institute, University of Florida College of Medicine, Gainesville, FL 32610, USA

Abstract

Recombinant Adeno-Associated Virus has become a promising vector for gene therapy and the treatment of neurological disorders. The fact that AAV has not been associated with any pathological condition in humans and that the recombinant vector displays widespread cell tropism and stable expression in different tissues have made AAV a favored gene transfer system in recent years. This chapter discusses different aspects of the biology of AAV and the properties of this viral vector in the nervous system. Finally, factors that are important for the safe use of this gene delivery system into humans are discussed.

Keywords: AAV, serotype, gene delivery, nervous system, brain, spinal cord, Tropism, neurological disorder.

Contents

1. Introduction 17
2. Biology of AAV 18
   2.1. AAV life cycle and genome structure 18
   2.2. AAV serotypes 19
3. AAV Gene transfer into the nervous system 19
   3.1. Transduction properties of different serotypes in the nervous system 19
   3.2. Methods of delivery 20
   3.3. Persistence of Transgene expression 21
4. Safety Considerations 22
   4.1. Transcriptional regulation of transgene expression 22
   4.2. Brain immune responses to rAAV 23
5. Conclusion 23
References 24

1. INTRODUCTION

Gene therapy for disorders of the central nervous system (CNS) is a moderately invasive method for delivering large molecules inside the blood–brain barrier (BBB) as compared to mechanical devices such as pumps or reservoirs.

Although many investigators are attempting to develop non-neurosurgical methods to deliver genes to the CNS, at least in the near term, CNS gene therapy will require a neurosurgical procedure to deliver the gene therapy vector. Therefore, at this initial phase, gene therapy for CNS disorders focuses into localized delivery of a given transgene. In particular, disorders where it is beneficial to overexpress or reduce expression of a transgene in a particular anatomical region to the exclusion of global delivery are especially attractive targets. Other disorders that require transduction of the entire brain or spinal cord are at the moment challenges for which optimization of diffusion of the vector and transduction efficiencies are in the developmental stage.

Several factors regarding recent progress both in understanding the pathogenesis of CNS diseases and advances in gene therapy vectors suggest that gene therapies aimed at the human CNS will become a reality sooner rather than later. Indeed, already five Phase I clinical trials have been approved using recombinant adeno-associated viral vectors (rAAV) for treatment of neurological disorders [1]. The present chapter will review some of the properties that make rAAV an attractive gene transfer system for nervous system disorders, as well as recent developments in the optimization of transduction due to the characterization of new serotypes. Finally, some limitations and challenges in the use of this viral vector will be discussed.

2. BIOLOGY OF AAV

2.1. AAV life cycle and genome structure

Wild-type adeno-associated virus (wt-AAV) is a small (~20 nm) member of the parvoviridiae family [2]. Wt-AAV has a single-stranded linear DNA genome of 4.7 kb in size that contains two genes, rep and cap, both of which are necessary for its life cycle [3,4]. These two genes are flanked by two inverted terminal DNA repeats (ITR) that serve as origins of viral replication [5,6], integration into the genomic DNA, and rescue from the latent phase [7–9]. The rAAV has been stripped of both rep and cap genes, and these have been substituted by an expression cassette containing a eukaryotic promoter directing transgene expression, a kozak sequence, and a polyadenylation sequence. This eukaryotic transcriptional cassette is flanked by the ITRs which are necessary for packaging and second strand synthesis of the viral vector [10]. The genes necessary for vector production are provided in trans, by a helper transfection system [11,12].

One limitation of rAAV is its small genome size of 4.7 kb, which limits the use of rAAV for large genes. Grieger and Samulski have shown that as the total packaging size becomes larger than 5.3 kb, cell transduction decreases, presumably due to preferential degradation of the virion in the cell [13]. We have successfully packaged and transduced vectors 5.4 kb in size using rAAV5 (Nash and Burger, unpublished observations). One solution to overcome this problem is to clone the gene in two different rAAV vectors that contain the gene in tandem, with the 5′ part of the gene and an a half-intron-carrying splice donor site in the first vector, and the second part of the gene in the second vector linked to the half-intron-carrying splice acceptor site [14–16]. Another option is to create an active mini-gene as has been shown with the dystrophin gene, which is mutated in muscular dystrophy [17,18], and with the B-domain–deleted human factor VIII cDNA for hemophilia [19].

Wt-AAV requires a helper virus for replication, such as herpes simplex virus or adenovirus [20,21]. In the absence of helper virus, wt-AAV integrates into the host cell genome and goes into a latent phase [22,23]. Recent data suggest that it most likely persists as episomal DNA in the cell nucleus [24–27] and it can persist in this form for a long time. About 80% of the human population is seropositive for AAV2 [28] yet, AAV2 infection has not been associated with any pathological condition in humans. The fact that the virus cannot replicate in the absence of a helper virus and that AAV is not associated with any disease, makes rAAV a safe gene therapy viral vector.

2.2. AAV serotypes

Currently, there are over one hundred new serotypes and genomic variants of AAV from humans and non-human primates [29–31]. Until the discovery of this burgeoning number of serotypes within the last five years, serotype 2 was used exclusively for gene therapy applications. The discovery of new serotypes has been important for a couple of reasons. First, the newly identified serotypes have shown distinct tropisms and broader transgene distribution in various organs including the nervous system, resulting in more versatility in the cellular targets accessible to viral transduction [31]. Second, given the preexisting immunity to AAV2 in most of the human population, the use of new immunologically distinct serotypes allows to escape the adaptive immune response to AAV2, which has been shown to be detrimental to viral transduction [32,33].

Different AAV serotypes interact with distinct cell surface receptors, thus explaining the diverse tropisms [34–38]. The receptor for AAV2 has been characterized and is composed of the heparan sulfate proteoglycan receptor, and at least one co-receptor, the fibroblast growth factor receptor-1 (FGFr1) [34,35]. The receptor for AAV5 is the platelet-derived growth factor receptor (PDGFr) [38]. The receptors for other serotypes are only beginning to be elucidated [39].

3. AAV GENE TRANSFER INTO THE NERVOUS SYSTEM

3.1. Transduction properties of different serotypes in the nervous system

The first serotype to be studied in the nervous system was rAAV2. rAAV2 has been demonstrated to achieve stable gene transfer in neurons in the CNS

[40–45] although gene transfer has been observed at a significantly lower potency in some non-neuronal cell types [40–42,44,46–48]. AAV2 has been shown to efficiently express transgenes in the substantia nigra [40,44,49–54], septal area [55–58], and the globus pallidus [53,59], and slightly less robustly the neostriatum [53,60–62]. Similarly, in the peripheral nervous system (PNS), rAAV2 gene transfer has been demonstrated in the dorsal root ganglia (DRG) and peripheral axons [63,64]. rAAV2 weakly transduces the hippocampus [41,43,44,53,65–75].

A number of newly identified serotypes have been vectored and tested in different tissues [30]. To date, rAAV1, rAAV2, rAAV4, rAAV5, and rAAV6 transduction properties have been described in the nervous system [47,53,73,76–78], although more recently serotypes 7, 8, 9, and Rh10 have also been described [79,80,81]. Specifically, rAAV1 and rAAV5 have demonstrated a broader diffusion and cell tropism than rAAV2 in the CNS [47,53,54,73,76,77,82–85]. rAAV4 targets solely ependymal cells, rAAV1,2 and rAAV5 target both neurons and ependymal cells [47,83]. In the retina, rAAV2 and rAAV5 transduce retinal-pigmented epithelial cells (RPE) and photoreceptor cells, with higher transduction in the photoreceptor cells, whereas rAAV4 transduces RPE cells only [84].

rAAV1 and rAAV5 have been shown to be more effective than rAAV2 in striatum, hippocampus, globus pallidus, substantia nigra, spinal cord, and cerebellum [47,53,54,76, 83]. rAAV2 gene transfer ability in the hippocampus is limited to the hilar region, however, rAAV1 and rAAV5 target the entire hippocampus including the pyramidal neurons in the CA1-CA3 region [38,53,86,87]. rAAV1 has demonstrated to be more efficient than rAAV2 in the cat brain (where rAAV5 did not transduce neurons to any significant level) [73]. Other promising serotypes for global brain transduction include rAAV9 and rAAVrh10 [80,81], since these appear to show the broadest distribution in the CNS of all the serotypes studied to date.

3.2. Methods of delivery

For neurological disorders that require global brain transduction, one of the current challenges of gene therapy is the ability to transduce sufficient cells in the CNS to successfully treat these diseases. In order to amplify the ability to transduce large areas of the CNS, the idea of intracarotid, intraventricular, or intrathecal vector injection of the vector with or without hyperosmolar disruptions of the BBB have been attempted [88–92]. However, these types of injections invariably result in low efficiency transduction of the parenchyma, due most likely to: (1) inability to pass the ependymal wall of ventricle due the size of the vector, (2) existence of viral receptors in the ependymal lining, and/ or (3) dilution of the vector by either blood or CSF to unacceptably low multiplicities of infection.

The use of different capsid serotypes such as rAAV1, 5, 8 and the most promising in terms of spread rAAV9 and rAAVrh10 [80,81] might help to solve this problem. In addition, some additional methods to further increase distribution of rAAV transduction in the CNS using these new serotypes will further help obtain global transduction in the large human brain. Methods to broaden the spread of viral transduction have been developed. Infusion of rAAV2 with heparin, presumably to block one of the co-receptors for AAV binding has been reported to increase the distribution of rAAV transduction [93,94]. Similarly, basic fibroblast growth factor, the other AAV co-receptor has also been used [62]. Convection-enhanced delivery, a slow, longterm infusion method, may enhance the distribution of rAAV-mediated transduction in the brain [93]. Lastly, mannitol, a complex sugar that produces hyperosmolality in the brain, thereby shrinking the interstitial space and reversing the flow of CSF, has been shown to increase the transduction rate when added to the vector sample [95] or after systemic injection [96,97].

3.3. Persistence of transgene expression

AAV not only transfers genes to the CNS with high efficacy, but also is capable of stable long-term expression [98–100]. Early observations using rAAV2 demonstrated long-term transgene expression in the CNS [40–44,49]. rAAV also transduces neurons long term in the spinal cord [42,53] and cells in the PNS [64,101]. Long-term expression in DRG has been observed after injection of rAAV2 into the sciatic nerve or DRG that lasted for at least 8 months [64].

The long-term expression of rAAV-delivered genes has been exploited to study the correction of various nervous system-related diseases in rodent models. rAAV2 has been shown to correct central diabetes insipidus for at least 50 weeks post-injection [102]. rAAV2 expressing Brain-Derived Neurotrophic Factor (BDNF) has been demonstrated to give persistent transduction for up to 18 months [103]. In parkinsonian rat models, rAAV2 vectors expressing tyrosine hydroxylase (TH), aromatic-L-amino acid decarboxylase (AADC) and guanosine triphosphate cyclohydroxylase (GCH) successfully increased dopamine levels in striatum and improved rotational behavior, which persisted for at least 12 months after intrastriatal injection [104–106].

rAAV-mediated glial cell line-derived neurotrophic factor (GDNF) delivery has also been shown to be effective in several animal models of Parkinson's disease (PD) [50,107–109]. In rodents, the GDNF was expressed for at least 6 months [107] and in the marmoset rAAV-mediated GDNF was expressed for at least 17 weeks [109].

Different rAAV serotypes (rAAV1, 2 and 5) have also been effective in long-term (4–12 months) correction of mucopolysaccharidosis (MPS) type I, II and VII lysosomal storage diseases [73,110–112]. In transgenic mice of Canavan's disease, rAAV2-containing aspartoacylase remarkably reduced the spongiform degeneration observed in the thalamus up to 5 months post-injection [113].

Large animals have also been examined as a better model of human viral transduction. Stable GFP expression has been observed in both RPE and photoreceptor cells in the eyes of both dogs and cynomologus macaque for at least 6 months post-injection using rAAV2, rAAV4, and rAAV5 [84]. Subretinal injections of rAAV2 containing RPE65 into RPE65-/-dogs resulted in correction of eyesight in the dogs' eyes for at least 9 months post-injection [114]. In the treatment of Parkinson's disease, persistent behavioral recovery up to 10 months post-injection of rAAV2 vectors (encoding TH, AADC, and GCH) has been observed in 1-methyl-4-phenyl-1,2,3,6-tetrahydropyridine-treated monkeys [115]. (*Note added in Proof*: Daadi et al. [147] have recently reported stable gene expression of AAV2 encoding human AADC 3 years after gene transfer in a nonhuman primate model of Parkinson's disease.)

In conclusion, the different rAAVs appear to sustain transgene expression for great lengths of time that can last for the length of the lifetime of the animal with no obvert side effects.

4. SAFETY CONSIDERATIONS

4.1. Transcriptional regulation of transgene expression

Long-term expression is indispensable in the treatment of neurodegenerative disorders. However, constitutive transgene expression also raises potential clinical safety concerns. The ability to regulate the levels of transgene expression of a gene that is successfully expressed via *ex vivo* or *in vivo* gene transfer in the CNS will ultimately be an important issue for this field of study. Using the cellular regulatory regions for a given gene has been unfeasible in most cases. In general, the genomic sequences that regulate transcription of genes span thousands of nucleotides which small viral genomes cannot accommodate. In some instances, short DNA sequences regulating a given gene have been identified and have been used for gene therapy [116–119].

Although other schemes exist, the main focus of transgene regulation has been at the transcriptional level by engineering internal promoters that are dependent upon binding some external prodrug, either for onset or cessation of promoter activity. The only one example of a naturally regulated promoter is the hypoxia promoter which has been used in gene therapy to target cancer cells [120], ischemic hearts [121], and in retinal vascular disease [122] In terms of synthetic-regulated promoters, the most developed is the tetracycline-regulated promoter system [123]. This system, which can also include both an inducer and a repressor system [124], has been utilized successfully in both viral vectors [98,125] and transgenic animals [126, p. 591; 127; 128, pp. 409, 590]. Other promising transcriptional regulation candidate systems that are amenable to gene transfer vectors also exist, one of which uses a progestin-inducible system [129–131] and another which uses a rapamycindimerizing system [132–134], ecdysone [135]. These various systems have been used in the context of rAAV vectors and have been discussed in detail in [136,137].

The further development of these gene-regulation systems is important for the future clinical progression of gene therapy. It is obvious that, with the exception of gene therapy strategies designed to destroy tumors or in cases where the transgene is completely non-toxic, virtually all gene therapy applications will require the ability to externally regulate protein levels in the CNS for individual dosing and as a safety net in the event of side effects. To date, no clinical trials involve regulated expression of genes.

4.2. Brain immune responses to rAAV

Eighty percent of the human population is positive for antibody to AAV serotype 2 (AAV2), with 30–70% of those individuals having neutralizing antibodies as well [28,138–140]. A dogmatic view of the brain as immunoprivileged has been challenged by recent studies [141]. Injection of rAAV vectors into the striatum of naïve animals results in limited immune response to the injection damage [46]. On the other hand, re-administration of the same serotype and/or preimmunization with rAAV results in an increase in neutralizing antibodies, with differing reports in the resulting efficiency in transgene expression [32,46,94,142]. These studies demonstrate that rAAV is capable of eliciting an immune response even in the brain. Therefore, vector design and immune status of patients should be considered when designing clinical trials.

5. CONCLUSION

Viral vectors are the most advantageous gene-delivery vehicles currently available because their wild-type counterparts have evolved to deliver genetic material to the cell. Advances in the characterization of the life cycle of wild-type viruses have allowed the engineering of recombinant viral vectors that are deficient in replication functions but are capable of infecting the cell to introduce foreign genes. Because the majority of cells in the CNS are post-mitotic, vectors that are capable of gene delivery to quiescent cells are of the most interest, and AAV has this ability.

The fact that AAV is a non-pathogenic virus, the discovery of new serotypes and the mainstreamed purification methods that allow high titer, pure preparations, has made AAV an attractive vector for human gene therapy. At this moment, there are at least five Phase I clinical trials using rAAV to treat neurological disorders [1]. There are also numerous studies using rAAV to create animal models of neurological disorders, and a great number of different molecular approaches to treat neurological disorders using AAV (reviewed in [143]). The future of gene therapy will depend in addressing safety issues such as the immune responses to viral delivery and regulation of the transgene in the trial design. As we have seen [144–146], avoidable mishaps in human gene therapy clinical trials could derail this promising field and careful consideration of known potential caveats with rAAV gene transfer in the CNS might minimize the potential for undesired setbacks.

REFERENCES

[1] R. J. Mandel and C. Burger, Clinical trials in neurological disorders using AAV vectors: Promises and challenges, *Curr. Opin. Mol. Ther.*, 2004, **6**(5), 482–490.

[2] N. Muzyczka and K. I. Berns, Parvoviridae: the viruses and their replication, in *Fields Virology* (eds. D. M. Knipe and P. M. Howley), Williams and Wilkins New York, Lippincott, 2001, pp. 2327–2360.

[3] A. Srivastava, E. W. Lusby and K. I. Berns, Nucleotide sequence and organization of the adeno-associated virus 2 genome, *J. Virol.*, 1983, **45**(2), 555–564.

[4] K. I. Berns and R. M. Linden, The cryptic life style of adeno-associated virus, *Bioessays*, 1995, **17**(3), 237–245.

[5] W. W. Hauswirth and K. I. Berns, Origin and termination of adeno-associated virus DNA replication, *Virology*, 1977, **78**(2), 488–499.

[6] E. Lusby, K. H. Fife, *et al.*, Nucleotide sequence of the inverted terminal repetition in adeno-associated virus DNA, *J. Virol.*, 1980, **34**(2), 402–409.

[7] R. M. Linden, P. Ward, *et al.*, Site-specific integration by adeno-associated virus, *Proc. Natl. Acad. Sci. USA*, 1996, **93**(21), 11288–11294.

[8] R. M. Linden, E. Winocour, *et al.*, The recombination signals for adeno-associated virus site-specific integration, *Proc. Natl. Acad. Sci. USA*, 1996, **93**(15), 7966–7972.

[9] C. Balague, M. Kalla, *et al.*, Adeno-associated virus Rep78 protein and terminal repeats enhance integration of DNA sequences into the cellular genome, *J. Virol.*, 1997, **71**(4), 3299–3306.

[10] S. K. McLaughlin, P. Collis, *et al.*, Adeno-associated virus general transduction vectors: Analysis of proviral structures, *J. Virol.*, 1988, **62**(6), 1963–1973.

[11] L. Cao, Y. Liu, *et al.*, High-titer, wild-type free recombinant adeno-associated virus vector production using intron-containing helper plasmids, *J. Virol.*, 2000, **74**(24), 11456–11463.

[12] S. Zolotukhin, M. Potter, *et al.*, Production and purification of serotype 1, 2, and 5 recombinant adeno-associated viral vectors, *Methods*, 2002, **28**(2), 158–167.

[13] J. C. Grieger and R. J. Samulski, Packaging capacity of adeno-associated virus serotypes: Impact of larger genomes on infectivity and postentry steps, *J. Virol.*, 2005, **79**(15), 9933–9944.

[14] D. Duan, Y. Yue, *et al.*, A new dual-vector approach to enhance recombinant adeno-associated virus-mediated gene expression through intermolecular cis activation, *Nat. Med.*, 2000, **6**(5), 595–598.

[15] H. Nakai, T. A. Storm, *et al.*, Increasing the size of rAAV-mediated expression cassettes in vivo by intermolecular joining of two complementary vectors, *Nat. Biotechnol.*, 2000, **18**(5), 527–532.

[16] L. Sun, J. Li, *et al.*, Overcoming adeno-associated virus vector size limitation through viral DNA heterodimerization, *Nat. Med.*, 2000, **6**(5), 599–602.

[17] P. Leone, C. G. Janson, *et al.*, Aspartoacylase gene transfer to the mammalian central nervous system with therapeutic implications for Canavan disease, *Ann. Neurol.*, 2000, **48**(1), 27–38.

[18] L. Wang, R. Calcedo, T. C. Nichols, D. A. Bellinger, A. Dillow, I. M. Verma and J. M. Wilson, Sustained correction of disease in naive and AAV2-pretreated hemophilia B dogs: AAV2/8 mediated, liver-directed gene therapy, *Blood*, 2005, **105**(8), 3079–3086.

[19] Chao, L. Mao, *et al.*, Sustained expression of human factor VIII in mice using a parvovirus-based vector, *Blood*, 2000, **95**(5), 1594–1599.
[20] R. W. Atchison, B. C. Casto, *et al.*, Adenovirus-associated defective virus particles, *Science*, 1965, **149**, 754–756.
[21] R. M. Buller, J. E. Janik, *et al.*, Herpes simplex virus types 1 and 2 completely help adenovirus-associated virus replication, *J. Virol.*, 1981, **40**(1), 241–247.
[22] R. M. Kotin and K. I. Berns, Organization of adeno-associated virus DNA in latently infected Detroit 6 cells, *Virology*, 1989, **170**(2), 460–467.
[23] R. J. Samulski, X. Zhu, *et al.*, Targeted integration of adeno-associated virus (AAV) into human chromosome 19, *EMBO. J.*, **10**(12), 3941–3950.
[24] D. Duan, P. Sharma, *et al.*, Circular intermediates of recombinant adeno-associated virus have defined structural characteristics responsible for long-term episomal persistence in muscle tissue, *J. Virol.*, 1998, **72**(11), 8568–8577.
[25] D. Duan, P. Sharma, *et al.*, Formation of adeno-associated virus circular genomes is differentially regulated by adenovirus E4 ORF6 and E2a gene expression, *J. Virol.*, 1999, **73**(1), 161–169.
[26] M. A. Kay and H. Nakai, Looking into the safety of AAV vectors, *Nature*, 2003, **424**(6946), 251.
[27] B. C. Schnepp, R. L. Jensen *et al.*, Characterization of adeno-associated virus genomes isiolated from human tissues, *J. Virol.*, 2005, **79**(23), 14793–14803.
[28] K. Erles, P. Sebokova, *et al.*, Update on the prevalence of serum antibodies (lgG and lgM) to adeno-associated virus (AAV), *J. Med. Virol.*, 1999, **59**(3), 406–411.
[29] G. Gao, M. R. Alvira, *et al.*, Adeno-associated viruses undergo substantial evolution in primates during natural infections, *Proc. Natl. Acad. Sci. USA*, **100**(10), 6081–6086.
[30] D. Grimm, and M. A. Kay, From virus evolution to vector revolution: Use of naturally occurring serotypes of adeno-associated virus (AAV) as novel vectors for human gene therapy, *Curr. Gene Ther.*, 2003, **3**(4), 281–304.
[31] G. Gao, L. H. Vandenberghe, *et al.*, New recombinant serotypes of AAV vectors, *Curr. Gene Ther.*, 2005, **5**(3), 285–297.
[32] C. S. Peden, C. Burger, *et al.*, Circulating anti-wild-type adeno-associated virus type 2 (AAV2) antibodies inhibit recombinant AAV2 (rAAV2)-mediated, but not rAAV5-mediated, gene transfer in the brain, *J. Virol.*, 2004, **78**(12), 6344–6359.
[33] B. P. De, A. Heguy, *et al.*, High levels of persistent expression of alpha1-antitrypsin mediated by the nonhuman primate serotype rh.10 adeno-associated virus despite pre-existing immunity to common human adeno-associated viruses, *Mol. Ther.*, 2006, **13**(1), 67–76.
[34] C. Summerford and R. J. Samulski, Membrane-associated heparan sulfate proteoglycan is a receptor for adeno-associated virus type 2 virions, *J. Virol.*, 1998, **72**(2), 1438–1445.
[35] K. Qing, C. Mah, *et al.*, Human fibroblast growth factor receptor 1 is a co-receptor for infection by adeno-associated virus 2, *Nat. Med.*, 1999, **5**(1), 71–77.
[36] C. Summerford, J. S. Bartlett, *et al.*, AlphaVbeta5 integrin: A co-receptor for adeno-associated virus type 2 infection, *Nat. Med.*, 1999, **5**(1), 78–82.
[37] R. W. Walters, S. M. Yi, *et al.*, Binding of adeno-associated virus type 5 to 2,3-linked sialic acid is required for gene transfer, *J. Biol. Chem.*, 2001, **276**(23), 20610–20616.
[38] G. Di Pasquale, B. L. Davidson, C. S. Stein, I. Martins, D. Scudiero, A. Monks and J. A. Chiorini, Identification of PDGFR as a receptor for AAV-5 transduction, *Nat. Med.*, 2003, **9**(10), 1306–1312.
[39] M. P. Seiler, A. D. Miller, J. Zabner and C. L. Halbert, Adeno-associated virus types 5 and 6 use distinct receptors for cell entry, *Hum. Gene Ther.*, 2006, **17**(1), 10–19.
[40] M. G. Kaplitt, P. Leone, *et al.*, Long-term gene expression and phenotypic correction using adeno-associated virus vectors in the mammalian brain, *Nat. Genet.*, 1994, **8**(2), 148–154.

[41] T. J. McCown, X. Xiao, *et al.*, Differential and persistent expression patterns of CNS gene transfer by an adeno-associated virus (AAV) vector, *Brain Res.*, 1996, **713**(1–2), 99–107.
[42] A. L. Peel, S. Zolotukhin, *et al.*, Efficient transduction of green fluorescent protein in spinal cord neurons using adeno-associated virus vectors containing cell type-specific promoters, *Gene Ther.*, 1997, **4**(1), 16–24.
[43] J. S. Bartlett, R. J. Samulski, *et al.*, Selective and rapid uptake of adeno-associated virus type 2 in brain, *Hum. Gene Ther.*, 1998, **9**(8), 1181–1186.
[44] R. L. Klein, E. M. Meyer, *et al.*, Neuron-specific transduction in the rat septohippocampal or nigrostriatal pathway by recombinant adeno-associated virus vectors, *Exp. Neurol.*, **150**(2), 183–194.
[45] R. J. Mandel, K. G. Rendahl, *et al.*, Characterization of intrastriatal recombinant adeno-associated virus-mediated gene transfer of human tyrosine hydroxylase and human GTP-cyclohydrolase I in a rat model of Parkinson's disease, *J. Neurosci.*, 1998, **18**(11), 4271–4284.
[46] W. D. Lo, G. Qu, *et al.*, Adeno-associated virus-mediated gene transfer to the brain: Duration and modulation of expression, *Hum. Gene Ther.*, 1999, **10**(2), 201–213.
[47] B. L. Davidson, C. S. Stein, *et al.*, Recombinant adeno-associated virus type 2, 4, and 5 vectors: transduction of variant cell types and regions in the mammalian central nervous system, *Proc. Natl. Acad. Sci. USA*, 2000, **97**(7), 3428–3432.
[48] M. Cucchiarini, X. L. Ren, *et al.*, Selective gene expression in brain microglia mediated via adeno-associated virus type 2 and type 5 vectors, *Gene Ther.*, 2003, **10**(8), 657–667.
[49] R. J. Mandel, S. K. Spratt, *et al.*, Midbrain injection of recombinant adeno-associated virus encoding rat glial cell line-derived neurotrophic factor protects nigral neurons in a progressive 6-hydroxydopamine-induced degeneration model of Parkinson's disease in rats, *Proc. Natl. Acad. Sci. USA*, 1997, **94**(25), 14083–14088.
[50] A. Bjorklund, D. Kirik, *et al.*, Towards a neuroprotective gene therapy for Parkinson's disease: Use of adenovirus, AAV and lentivirus vectors for gene transfer of GDNF to the nigrostriatal system in the rat Parkinson model, *Brain Res.*, 2000, **886**(1–2), 82–98.
[51] S. Furler, J. C. Paterna, *et al.*, Recombinant AAV vectors containing the foot and mouth disease virus 2A sequence confer efficient bicistronic gene expression in cultured cells and rat substantia nigra neurons, *Gene Ther.*, 2001, **8**(11), 864–873.
[52] L. Wang, S. Muramatsu, *et al.*, Delayed delivery of AAV-GDNF prevents nigral neurodegeneration and promotes functional recovery in a rat model of Parkinson's disease, *Gene Ther.*, 2002, **9**(6), 381–389.
[53] C. Burger, O. S. Gorbatyuk, *et al.*, Recombinant AAV viral vectors pseudotyped with viral capsids from serotypes 1, 2, and 5 display differential efficiency and cell tropism after delivery to different regions of the central nervous system, *Mol. Ther.*, 2004, **10**(2), 302–317.
[54] J. C. Paterna, J. Feldon, *et al.*, Transduction profiles of recombinant adeno-associated virus vectors derived from serotypes 2 and 5 in the nigrostriatal system of rats, *J. Virol.*, 2004, **78**(13), 6808–68017.
[55] R. J. Mandel, F. H. Gage, *et al.*, Nerve growth factor expressed in the medial septum following in vivo gene delivery using a recombinant adeno-associated viral vector protects cholinergic neurons from fimbria-fornix lesion-induced degeneration, *Exp. Neurol.*, 1999, **155**(1), 59–64.
[56] R. L. Klein, A. C. Hirko, *et al.*, NGF gene transfer to intrinsic basal forebrain neurons increases cholinergic cell size and protects from age-related, spatial memory deficits in middle-aged rats, *Brain Res.*, 2000, **875**(1–2), 144–151.
[57] B. K. Kaspar, B. Vissel, *et al.*, Adeno-associated virus effectively mediates conditional gene modification in the brain, *Proc. Natl. Acad. Sci. USA*, 2002, **99**(4), 2320–2325.

[58] R. Landgraf, E. Frank, *et al.*, Viral vector-mediated gene transfer of the vole V1a vasopressin receptor in the rat septum: Improved social discrimination and active social behaviour, *Eur. J. Neurosci.*, 2003, **18**(2), 403–411.
[59] L. Tenenbaum, F. Jurysta, *et al.*, Tropism of AAV-2 vectors for neurons of the globus pallidus, *Neuroreport*, 2000, **11**(10), 2277–2283.
[60] D. S. Fan, M. Ogawa, *et al.*, Behavioral recovery in 6-hydroxydopamine-lesioned rats by cotransduction of striatum with tyrosine hydroxylase and aromatic L-amino acid decarboxylase genes using two separate adeno-associated virus vectors, *Hum. Gene Ther.*, 1998, **9**(17), 2527–2535.
[61] B. Dass, M. M. Iravani, *et al.*, Sonic hedgehog delivered by an adeno-associated virus protects dopaminergic neurones against 6-OHDA toxicity in the rat, *J. Neural. Transm.*, 2004.
[62] P. Hadaczek, H. Mirek, *et al.*, Basic fibroblast growth factor enhances transduction, distribution, and axonal transport of adeno-associated virus type 2 vector in rat brain, *Hum. Gene Ther.*, 2004, **15**(5), 469–479.
[63] J. Fleming, S. L. Ginn, *et al.*, Adeno-associated virus and lentivirus vectors mediate efficient and sustained transduction of cultured mouse and human dorsal root ganglia sensory neurons, *Hum. Gene Ther.*, 2001, **12**(1), 77–86.
[64] Y. Xu, Y. Gu, *et al.*, Efficiencies of transgene expression in nociceptive neurons through different routes of delivery of adeno-associated viral vectors, *Hum. Gene Ther.*, 2003, **14**(9), 897–906.
[65] A. Freese, M. G. Kaplitt, *et al.*, Direct gene transfer into human epileptogenic hippocampal tissue with an adeno-associated virus vector: Implications for a gene therapy approach to epilepsy, *Epilepsia*, 1997, **38**(7), 759–766.
[66] Shimazaki, M. Urabe, *et al.*, Adeno-associated virus vector-mediated bcl-2 gene transfer into post-ischemic gerbil brain in vivo: Prospects for gene therapy of ischemia-induced neuronal death, *Gene Ther.*, 2000, **7**(14), 1244–1249.
[67] M. U. Ehrengruber, S. Hennou, *et al.*, Gene transfer into neurons from hippocampal slices: Comparison of recombinant Semliki Forest virus, adenovirus, adeno-associated virus, lentivirus, and measles virus, *Mol. Cell. Neurosci.*, 2001, **17**(5), 855–871.
[68] B. K. Kaspar, D. Erickson, *et al.*, Targeted retrograde gene delivery for neuronal protection, *Mol. Ther.*, 2002, **5**(1), 50–56.
[69] R. L. Klein, M. E. Hamby, *et al.*, Measurements of vector-derived neurotrophic factor and green fluorescent protein levels in the brain, *Methods*, 2002, **28**(2), 286–292.
[70] M. A. Passini, E. B. Lee, *et al.*, Distribution of a lysosomal enzyme in the adult brain by axonal transport and by cells of the rostral migratory stream, *J. Neurosci.*, 2002, **22**(15), 6437–6446.
[71] R. P. Haberman, R. J. Samulski, *et al.*, Attenuation of seizures and neuronal death by adeno-associated virus vector galanin expression and secretion, *Nat. Med.*, 2003, **9**(8), 1076–1080.
[72] E. J. Lin, C. Richichi, *et al.*, Recombinant AAV-mediated expression of galanin in rat hippocampus suppresses seizure development, *Eur. J. Neurosci.*, 2003, **18**(7), 2087–2092.
[73] M. A. Passini, D. J. Watson, *et al.*, Intraventricular brain injection of adeno-associated virus type 1 (AAV1) in neonatal mice results in complementary patterns of neuronal transduction to AAV2 and total long-term correction of storage lesions in the brains of \beta;-glucuronidase-deficient mice, *J. Virol.*, 2003, **77**(12), 7034–7040.
[74] T. E. Scammell, E. Arrigoni, *et al.*, Focal deletion of the adenosine A1 receptor in adult mice using an adeno-associated viral vector, *J. Neurosci.*, 2003, **23**(13), 5762–5770.
[75] B. Y. Ahmed, S. Chakravarthy, *et al.*, Efficient delivery of Cre-recombinase to neurons in vivo and stable transduction of neurons using adeno-associated and lentiviral vectors, *BMC Neurosci.*, 2004, **5**(1), 4.

[76] J. M. Alisky, S. M. Hughes, *et al.*, Transduction of murine cerebellar neurons with recombinant FIV and AAV5 vectors, *Neuroreport*, 2000, **11**(12), 2669–2673.
[77] G. S. Yang, M. Schmidt, *et al.*, Virus-mediated transduction of murine retina with adeno-associated virus: Effects of viral capsid and genome size, *J. Virol.*, 2002, **76**(15), 7651–7660.
[78] C. H. Vite, M. A. Passini, *et al.*, Adeno-associated virus vector-mediated transduction in the cat brain, *Gene Ther.*, 2003, **10**(22), 1874–1881.
[79] R. L. Klein, R. D. Dayton, N. J. Leidenheimer, K. Jansen, T. E. Golde and R. M. Zweig, Efficient neuronal gene transfer with AAV8 leads to neurotoxic levels of \tau; or green fluorescent proteins, *Mol. Ther.*, 2006, **13**(3), 517–527.
[80] C. N. Cearley and J. H. Wolfe, Transduction characteristics of adeno-associated virus vectors expressing cap serotypes 7, 8, 9, and Rh10 in the mouse brain, *Mol. Ther.*, 2006, **13**(3), 528–537.
[81] D. Sondhi, D. A. Peterson, E. Vassallo, C. T. Sanders, J. A. Stratton, N. R. Hackett, G. Gao, J. M.Wilson and R.G. Crystal, *Quantitative Comparison of AAV Serotypes AAV2, AAV5, and AAVrh.10 Efficiency for CNS Gene Therapy following Intracranial Gene Delivery*, American Society for Gene Therapy, St. Louis, MO, 2005.
[82] A. Auricchio, G. Kobinger, *et al.*, Exchange of surface proteins impacts on viral vector cellular specificity and transduction characteristics: The retina as a model, *Hum. Mol. Genet.*, 2001, **10**(26), 3075–3081.
[83] C. Wang, C. M. Wang, *et al.*, Recombinant AAV serotype 1 transduction efficiency and tropism in the murine brain, *Gene Ther.*, 2003, **10**(17), 1528–1534.
[84] M. Weber, J. Rabinowitz, *et al.*, Recombinant adeno-associated virus serotype 4 mediates unique and exclusive long-term transduction of retinal pigmented epithelium in rat, dog, and nonhuman primate after subretinal delivery, *Mol. Ther.*, 2003, **7**(6), 774–781.
[85] D. Lin, C. R. Fantz, *et al.*, AAV2/5 vector expressing galactocerebrosidase ameliorates CNS disease in the murine model of globoid-cell leukodystrophy more efficiently than AAV2, *Mol. Ther.*, 2005, **12**(3), 422–430.
[86] C. Richichi, E. J. Lin, *et al.*, Anticonvulsant and antiepileptogenic effects mediated by adeno-associated virus vector neuropeptide y expression in the rat hippocampus, *J. Neurosci.*, 2004, **24**(12), 3051–3059.
[87] M. Klugmann, C. Wymond Symes, *et al.*, AAV-mediated hippocampal expression of short and long homer 1 proteins differentially affect cognition and seizure activity in adult rats, *Mol. Cell. Neurosci.*, 2005, **28**(2), 347–360.
[88] A. L. Betz, P. Shakui, *et al.*, Gene transfer to rodent brain with recombinant adenoviral vectors: Effects of infusion parameters, infectious titer, and virus concentration on transduction volume, *Exp. Neurol.*, **150**(1), 136–142.
[89] A. J. Mannes, R. M. Caudle, *et al.*, Adenoviral gene transfer to spinal-cord neurons: Intrathecal vs. intraparenchymal administration, *Brain Res.*, 1998, **793**(1–2), 1–6.
[90] M. J. Driesse, J. M. Kros, *et al.*, Distribution of recombinant adenovirus in the cerebrospinal fluid of nonhuman primates, *Hum. Gene Ther.*, **10**(14), 2347–2354.
[91] M. J. Driesse, M. C. Esandi, *et al.*, Intra-CSF administered recombinant adenovirus causes an immune response-mediated toxicity, *Gene Ther.*, 2000, **7**(16), 1401–1409.
[92] Lundberg, S. J. Jungles, *et al.*, Direct delivery of leptin to the hypothalamus using recombinant adeno-associated virus vectors results in increased therapeutic efficacy, *Nat. Biotechnol.*, 2001, **19**(2), 169–172.
[93] J. B. Nguyen, R. Sanchez-Pernaute, *et al.*, Convection-enhanced delivery of AAV-2 combined with heparin increases TK gene transfer in the rat brain, *Neuroreport*, 2001, **12**(9), 1961–1964.
[94] M. Y. Mastakov, K. Baer, *et al.*, Recombinant adeno-associated virus serotypes 2- and 5-mediated gene transfer in the mammalian brain: Quantitative analysis of heparin co-infusion, *Mol. Ther.*, 2002, **5**(4), 371–380.

[95] M. Y. Mastakov, K. Baer, *et al.*, Combined injection of rAAV with mannitol enhances gene expression in the rat brain, *Mol. Ther.*, 2001, **3**(2), 225–232.

[96] H. Fu, J. Muenzer, *et al.*, Self-complementary adeno-associated virus serotype 2 vector: Global distribution and broad dispersion of AAV-mediated transgene expression in mouse brain, *Mol. Ther.*, 2003, **8**(6), 911–917.

[97] C. Burger, F. N. Nguyen, *et al.*, Systemic mannitol-induced hyperosmolality amplifies rAAV2-mediated striatal transduction to a greater extent than local co-infusion, *Mol. Ther.*, 2005, **11**(2), 327–331.

[98] K. G. Rendahl, S. E. Leff, *et al.*, Regulation of gene expression in vivo following transduction by two separate rAAV vectors, *Nat. Biotechnol.*, 1998, **16**(8), 757–761.

[99] R. I. Owen, A. P. Lewin, *et al.*, Recombinant adeno-associated virus vector-based gene transfer for defects in oxidative metabolism, *Hum. Gene Ther.*, 2000, **11**(15), 2067–2078.

[100] R. T. Owen, R. J. Mandel, *et al.*, Gene therapy for pyruvate dehydrogenase E1alpha deficiency using recombinant adeno-associated virus 2 (rAAV2) vectors, *Mol. Ther.*, 2002, **6**(3), 394–399.

[101] V. N. Martinov, I. Sefland, *et al.*, Targeting functional subtypes of spinal motoneurons and skeletal muscle fibers in vivo by intramuscular injection of adenoviral and adeno-associated viral vectors, *Anat. Embryol. (Berl)*, 2002, **205**(3), 215–221.

[102] J. Ideno, H. Mizukami, *et al.*, Persistent phenotypic correction of central diabetes insipidus using adeno-associated virus vector expressing arginine-vasopressin in Brattleboro rats, *Mol. Ther.*, 2003, **8**(6), 895–902.

[103] M. J. Ruitenberg, B. Blits, *et al.*, Adeno-associated viral vector-mediated gene transfer of brain-derived neurotrophic factor reverses atrophy of rubrospinal neurons following both acute and chronic spinal cord injury, *Neurobiol. Dis.*, 2004, **15**(2), 394–406.

[104] R. J. Mandel, K. G. Rendahl, *et al.*, Progress in direct striatal delivery of L-dopa via gene therapy for treatment of Parkinson's disease using recombinant adeno-associated viral vectors, *Exp. Neurol.*, 1999, **159**(1), 47–64.

[105] Y. Shen, S. I. Muramatsu, *et al.*, Triple transduction with adeno-associated virus vectors expressing tyrosine hydroxylase, aromatic-L-amino-acid decarboxylase, and GTP cyclohydrolase I for gene therapy of Parkinson's disease, *Hum. Gene Ther.*, 2000, **11**(11), 1509–1519.

[106] D. Kirik, B. Georgievska, *et al.*, Reversal of motor impairments in parkinsonian rats by continuous intrastriatal delivery of L-dopa using rAAV-mediated gene transfer, *Proc. Natl. Acad. Sci. USA*, 2002, **99**(7), 4708–4713.

[107] D. Kirik, C. Rosenblad *et al.*, Characterization of behavioral and neurodegenerative changes following partial lesions of the nigrostriatal dopamine system induced by intrastriatal 6-hydroxydopamine in the rat, *Exp. Neurol.*, 1998, **152**(2), 259–277.

[108] A. Eslamboli, R. M. Cummings, *et al.*, Recombinant adeno-associated viral vector (rAAV) delivery of GDNF provides protection against 6-OHDA lesion in the common marmoset monkey (*Callithrix jacchus*), *Exp. Neurol.*, 2003, **184**(1), 536–548.

[109] A. Eslamboli, B. Georgievska, *et al.*, Continuous low-level glial cell line-derived neurotrophic factor delivery using recombinant adeno-associated viral vectors provides neuroprotection and induces behavioral recovery in a primate model of Parkinson's disease, *J. Neurosci.*, 2005, **25**(4), 769–777.

[110] A. Bosch, E. Perret, *et al.*, Long-term and significant correction of brain lesions in adult mucopolysaccharidosis type VII mice using recombinant AAV vectors, *Mol. Ther.*, 2000, **1**(1), 63–70.

[111] H. Fu, R. J. Samulski, *et al.*, Neurological correction of lysosomal storage in a mucopolysaccharidosis IIIB mouse model by adeno-associated virus-mediated gene delivery, *Mol. Ther.*, 2002, **5**(1), 42–49.

[112] N. Desmaris, L. Verot, *et al.*, Prevention of neuropathology in the mouse model of Hurler syndrome, *Ann. Neurol.*, **56**(1), 68–76.

[113] R. Matalon, S. Surendran *et al.*, Adeno-associated virus-mediated aspartoacylase gene transfer to the brain of knockout mouse for canavan disease, *Mol. Ther.*, 2003, **7**(5, Pt 1), 580–587.
[114] K. Narfstrom, M. L. Katz, *et al.*, Functional and structural recovery of the retina after gene therapy in the RPE65 null mutation dog, *Invest. Ophthalmol. Vis. Sci.*, **44**(4), 1663–1672.
[115] S. Muramatsu, K. Fujimoto, *et al.*, Behavioral recovery in a primate model of Parkinson's disease by triple transduction of striatal cells with adeno-associated viral vectors expressing dopamine-synthesizing enzymes, *Hum. Gene Ther.*, **13**(3), 345–354.
[116] N. Miller and J. Whelan, Progress in transcriptionally targeted and regulatable vectors for genetic therapy, *Hum. Gene Ther.*, 1997, **8**(7), 803–815.
[117] A. S. Lewin, K. A. Drenser, *et al.*, Ribozyme rescue of photoreceptor cells in a transgenic rat model of autosomal dominant retinitis pigmentosa, *Nat. Med.*, 1998, **4**(8), 967–971.
[118] S. Kugler, P. Lingor, *et al.*, Differential transgene expression in brain cells in vivo and in vitro from AAV-2 vectors with small transcriptional control units, *Virology*, 2003, **311**(1), 89–95.
[119] H. Sadeghi and M. M. Hitt, Transcriptionally targeted adenovirus vectors, *Curr. Gene Ther.*, 2005, **5**(4), 411–427.
[120] H. Ruan, H. Su, *et al.*, A hypoxia-regulated adeno-associated virus vector for cancer-specific gene therapy, *Neoplasia*, 2001, **3**(3), 255–263.
[121] N. Iwata, H. Mizukami, *et al.*, Presynaptic localization of neprilysin contributes to efficient clearance of amyloid-beta peptide in mouse brain, *J. Neurosci.*, 2004, **24**(4), 991–998.
[122] J. W. Bainbridge, A. Mistry, *et al.*, Hypoxia-regulated transgene expression in experimental retinal and choroidal neovascularization, *Gene Ther.* 2003, **10**(12), 1049–1054.
[123] M. Gossen and H. Bujard, Tight control of gene expression in mammalian cells by tetracycline-responsive promoters, *Proc. Natl. Acad. Sci. USA*, 1992, **89**(12), 5547–5551.
[124] I. M. Mansuy and H. Bujard,Tetracycline-regulated gene expression in the brain, *Curr. Opin. Neurobiol.*, 2000, **10**(5), 593–596.
[125] T. Kafri, H. van Praag, *et al.*, Lentiviral vectors: Regulated gene expression, *Mol. Ther.*, 2000, **1**(6), 516–521.
[126] P. Tremblay, Z. Meiner, *et al.*, Doxycycline control of prion protein transgene expression modulates prion disease in mice, *Proc. Natl. Acad. Sci. USA*, 1998, **95**(21) 12580–12585.
[127] H. Bujard, Controlling genes with tetracyclines, *J. Gene Med.*, 1999, **1**, 372–374.
[128] U. Baron and H. Bujard, Tet repressor-based system for regulated gene expression in eukaryotic cells: principles and advances, *Methods Enzymol.*, 2000, **327**, 401–421.
[129] Y. Wang, B. W. O'Malley Jr. *et al.*, A regulatory system for use in gene transfer, *Proc. Natl. Acad. Sci. USA*, 1994, **91**(17), 8180–8184.
[130] D. M. Harvey and C. T. Caskey, Inducible control of gene expression: Prospects for gene therapy, *Curr. Opin. Chem. Biol.*, 1998, **2**(4), 512–518.
[131] T. Oligino, P. L. Poliani, *et al.*, Drug inducible transgene expression in brain using a herpes simplex virus vector, *Gene Ther.*, 1998, **5**(4), 491–496.
[132] V. M. Rivera, T. Clackson, *et al.*, A humanized system for pharmacologic control of gene expression, *Nat. Med.*, 1996, **2**(9), 1028–1032.
[133] J. F. Amara, T. Clackson, *et al.*, A versatile synthetic dimerizer for the regulation of protein–protein interactions, *Proc. Natl. Acad. Sci. USA*, 1997, **94**(20), 10618–10623.
[134] X. Ye, V. M. Rivera, *et al.*, Regulated delivery of therapeutic proteins after in vivo somatic cell gene transfer, *Science*, 1999, **283**(5398), 88–91.
[135] H. Okada, K. Miyamura, *et al.*, Gene therapy against an experimental glioma using adeno-associated virus vectors, *Gene Ther.*, 1996, **3**(11), 957–964.

[136] R. Haberman, H. Criswell, *et al.*, Therapeutic liabilities of in vivo viral vector tropism: Adeno-associated virus vectors, NMDAR1 antisense, and focal seizure sensitivity, *Mol. Ther.*, 2002, **6**(4), 495–500.
[137] S. Goverdhana, M. Puntel, *et al.*, Regulatable gene expression systems for gene therapy applications: Progress and future challenges, *Mol. Ther.*, 2005, **12**(2), 189–211.
[138] N. R. Blacklow, M. D. Hoggan, *et al.*, Serologic evidence for human infection with adenovirus-associated viruses, *J. Natl. Cancer Inst.*, 1968, **40**(2), 319–327.
[139] N. R. Blacklow, M. D. Hoggan, *et al.*, A seroepidemiologic study of adenovirus-associated virus infection in infants and children, *Am. J. Epidemiol.*, 1971, **94**(4), 359–366.
[140] N. Chirmule, K. Propert, *et al.*, Immune responses to adenovirus and adeno-associated virus in humans, *Gene Ther.*, 1999, **6**(9), 1574–1583.
[141] J. P. Stables, E. H. Bertram, *et al.*, Models for epilepsy and epileptogenesis: Report from the NIH workshop, Bethesda, Maryland, *Epilepsia*, 2002, **43**(11), 1410–1420.
[142] L. M. Sanftner, B. M. Suzuki *et al.*, Striatal delivery of rAAV-hAADC to rats with pre-existing immunity to AAV, *Mol. Ther.*, 2004, **9**(3), 403–409.
[143] R. J. Mandel, F. P. Manfredsson, K. D. Foust, A. Rising, S. Reimsnider, K. Nash and C. Burger, Recombinant adeno-associated viral vectors as therapeutic agents to treat neurological disorders, *Mol. Ther.*, 2006, **13**(3), 463–483.
[144] E. Marshall, Gene therapy death prompts review of adenovirus vector, *Science*, **286**(5448), 2244–2245.
[145] E. Marshall, Improving gene therapy's tool kit, *Science*, 2000, **288**(5468), 953.
[146] N. Somia, and I. M. Verma, Gene therapy: Trials and tribulations, *Nat. Rev. Genet.*, 2000, **1**(2), 91–99.
[147] M. M. Daadi, P. Pivirotto, J. Bringas, J. Cunnigham, J. Forsayeth, J. Eberling and K. S. Bankiewicz, Distribution of AAV2-hAADC-transduced cells after 3 years in Parkinsonian monkeys. *Neuroreport*, 2006, **17**(2), 201–204.

Controlling Adenoviral Gene Transfer in Heart by Catheter-Based Coronary Perfusion

J. Michael O'Donnell and E. Douglas Lewandowski

Program in Integrative Cardiac Metabolism, Center for Cardiovascular Research, University of Illinois at Chicago, College of Medicine, Chicago, IL 60612, USA

Abstract

Myocardial gene therapy represents a promising approach for the treatment of inherited heart diseases, cardiomyopathies, and congestive heart failure. Extensive work has demonstrated that recombinant adenoviral vectors can efficiently transduce cardiomyocytes *in vivo*. However, targeting gene expression to the heart and controlling the level of transfer has been a challenge. In this study, we describe a catheter-based method for gene delivery that results in 58% efficient gene transfer to rat heart with minimal infection of peripheral organs. The heart is completely isolated *in vivo* during delivery, and unsequestered virus is flushed from the heart prior to releasing the cross-clamp. We compare the results to existing methods and highlight promising new strategies.

Keywords: adenovirus, heart, gene therapy, *in vivo*.

Contents

1. Introduction 34
2. Methods 35
   2.1. Large-scale adenovirus preparation 35
      2.1.1. Protocol 1. Gene transfer by complete isolation 36
      2.1.2. Protocol 2. Gene transfer by partial isolation 37
      2.1.3. Assessing efficiency of gene transfer 37
      2.1.4. Assessing specificity of gene transfer 39
3. Results 40
   3.1. Survival 40
   3.2. Efficiency 40
   3.3. Specificity 40
4. Discussion 41
   4.1. Efficiency – Gene transfer to the whole heart 42
   4.2. Specificity – Targeting gene transfer to the heart 43
   4.3. Minimally invasive gene transfer – Promising technologies 43
5. Conclusion 44
Acknowledgements 44
References 44

1. INTRODUCTION

More than a century ago the Wright brothers received wide acclaim for being the first to demonstrate controlled flight. Historians are careful to recognize that several others demonstrated flight earlier, whereas the brother's fame was based on controlling and sustaining that flight. Today, the endeavor has changed, but the goal is the same. Several groups have demonstrated gene transfer to the heart *in vivo*, but controlling and reproducing the procedure has been a challenge. Controlling gene transfer *in vivo* presents three major obstacles: the approach needs to be (i) cardiac specific, (ii) highly efficient, and (iii) minimally invasive. At this time, there has been no report of an approach that meets all three criteria. In this study, we will describe a technique advanced in our laboratory that meets the first two criteria. We will compare the results to existing methods and highlight promising new strategies.

Several creative approaches have been developed to transfer exogenous genes to heart *in vivo* using adenoviral vectors. These techniques include (i) direct injection of the virus into the heart tissue [1], (ii) catheter-based coronary perfusion [2–4], (iii) cardiopulmonary bypass [5], (iv) transplantation [6], and (v) ultrasound micro-bubble destruction [7]. The focus of this report is on the catheter-based coronary perfusion of rodent heart with the recombinant adenovirus (rAd).

The catheter-based delivery approach was first described by Barr in 1994 [8]. A catheter was inserted into the right carotid artery and advanced into the coronary ostia before injecting 1 ml Ad.cmv.lacZ into the coronary artery for 1 min. A cross section of heart stain for lacZ expression showed myocytes expressing the exogenous gene. However, infection was sparse and efficiency was <1%. An important advancement to the approach was presented by Hajjar's laboratory in 1998 [4]. The viral solution was injected into the aortic root after first cross-clamping the aorta and pulmonary artery. This enabled both a high perfusion pressure and longer incubation periods prior to releasing the cross-clamp. As shown by our laboratory and several others [3,9–11], this resulted in 3–5% infection of the whole heart. This was a significant improvement over Barr's earlier work. Subsequent refinements to this basic approach have since yielded efficiencies as high as 40–80% in hamster [3], mouse [12,13], and rat [9,11]. Unfortunately, while the technique is clearly efficient at transferring exogenous genes to the whole heart *in vivo*, the advances fail to be both cardiac specific and minimally invasive.

It is still not clear how critical cardiac specificity is to gene transfer. Following coronary perfusion of the heart with the adenovirus, unsequestered virus accumulates in the liver and spleen [3,4,8,9]. These organs show very high expression of the genes targeted for the heart. It is not known if non-target gene expression in the liver and spleen are deleterious or influence the function of other organs. Could the many changes reported for the heart following gene transfer simply be a secondary response to liver infection? In this study we present an alternative. We target gene transfer to cardiomyocytes by isolating the heart *in vivo*. In so doing, we are able to control the many factors required to achieve highly efficient gene transfer *in vivo*. We are also able to minimize infection of peripheral tissue by flushing unsequestered virus out of the heart prior to releasing the cross-clamp.

2. METHODS

2.1. Large-scale adenovirus preparation

The concentration of virus delivered to the heart is a critical factor affecting gene transfer efficiency [14]. Viral stocks were amplified according to standard procedures with the following modifications [15,16]. The viral vector, Ad5.cmv.LacZ, that carries the *Escherichia coli* β-galactosidase gene under the control of the cytomegalovirus promoter (CMV) was purchased from QBioGene (Montreal, Canada). 10 μl of virus (1×10^{12} viral particles/ml) was used to infect two plates (p150) of HEK293 cells at 90% confluency. The growth medium (MEM) was supplemented with 5% horse serum (HS, Gibco) and antibiotics (penicillin/streptomycin/ampicillin). After 50% of the cells had detached (2–3 days), the cells were harvested and centrifuged at 2000 rpm for 10 min at 4°C. The supernatant was saved and the cell pellet was resuspended in MEM (w/o HS) and frozen/thawed/vortexed three times. The cells were again centrifuged, the supernatant combined with the first, and the pellet discarded. Enough medium (MEM/5% HS P/S/A) was added to the supernatant to infect eight plates (p150) of HEK293 cells. Later, these cells were harvested and the virus isolated as before. This second amplification was used to infect an additional 70 plates.

The final amplification was harvested, centrifuged, and the cell pellets were all combined into 14 ml sterile PBS. The cell culture medium from this final amplification was discarded (though kits are now available to isolate the virus from the medium). The cells were frozen/thawed/vortexed four times before centrifugation at 3000 rpm for 10 min. The supernatant was filtered (5.0 μm filter unit, Millipore) and added to a cesium chloride column (1.2 and 1.45 g/ml layers; 40 ml centrifuge tube) for isolation and purification by ultracentrifugation at 22,500 rpm at 4°C overnight. The viral band was aspirated from the column, filtered (5.0 μm filter), and dialyzed overnight (10 mM Tris, 1 mM $MgCl_2$, 10% glycerol). A final dialysis in glycerol-free PBS was performed before storing the stock in 0.5 ml samples and frozen at –80°C. Typical amplifications resulted in 2–3 ml of virus at 1×10^{12} viral particles/ml. The titers were measured by UV and used within 1–2 weeks.

2.1.1. Protocol 1. Gene transfer by complete isolation

The protocols were approved by the Animal Care Policies and Procedures Committee at the UIC (IACUC accredited), and animals used were maintained in accordance with the Guide for the Care and Use of Laboratory Animals (National Research Council, revised 1996). The details of the method have been described in our earlier Gene Therapy paper [9], and they are based on modifications and advances made to the earlier procedure described by Hajjar for gene transfer following partial isolation [4]. Figure 1 illustrates the approach. In brief, Sprague Dawley rats (300 gm) were intubated and anesthetized by mechanical ventilation with isoflurane (1.5% v/v) in 100% oxygen. The rats were placed on an ice pack and core body temperature was monitored. The chest was entered through a right thoracotomy between the second and third intercostals space. The pericardial sac was cut, pulled aside, and two sutures were incised at the apex of the heart for later use.

Once the core body temperature was below 30°C, the heart was externalized from the chest by pulling on the sutures. A 20-gauge catheter was inserted into the left ventricle at the apex of the heart. The catheter was advanced to the aortic root and the tip of the catheter was confirmed through the aortic wall. The base of the catheter was tied to the sutures positioned previously at the apex. All vessels leading to/from the heart were cross-clamped simultaneously with a single clamp. Care was taken to ensure that the clamp was distal to the catheter tip in the aortic root. The anesthetic was lowered to 0.5% isoflurane. Well-oxygenated, calcium-free Tyrode solution was pumped

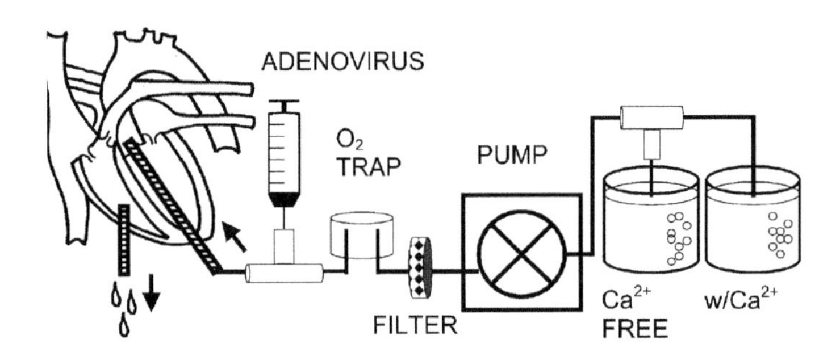

Fig. 1. Perfusion schematic. All vessels leading to/from the heart were cross-clamped simultaneously with a single clamp. A catheter positioned in the aortic root delivered calcium-free Tyrode solution, and a second catheter in the right ventricle provided a path for coronary efflux. After a 7.5 min permeability treatment period, perfusion was stopped and a single bolus of adenovirus (0.4 ml at 1×10^{12} viral particles/ml) was injected into the aortic root. After 90 s, the perfusion was continued with Krebs buffer containing normal levels of calcium. Once the heart regained a spontaneous heartbeat, the cross-clamp was removed.

through the catheter, and a second catheter (24 gauge) positioned in the right ventricle provided a path for coronary effluent. Flow was maintained at 3–5 ml/min (40 ± 10 mmHg) for 7.5 min. This period served to enhance the endothelial permeability based on a calcium-free strategy previously reported for *ex vivo* gene transfer [17]. Importantly, to ensure proper perfusion of the coronaries during the entire protocol, care was taken not to pull or compress the heart.

The peristaltic pump was stopped at the end of the permeability enhancement period, and a single bolus of 0.4 ml virus (10^{12} viral particles/ml) was injected into the aortic root within 2 s (peak pressure 400 ± 100 mmHg). Sham-operated controls were injected with 0.4 ml PBS. Following a 90 s incubation period, the perfusate supply line was switched to well-oxygenated Krebs' buffer solution containing normal levels of calcium (1.5 mM). Once the heart had regained a spontaneous heartbeat, the catheters and cross-clamp were removed. Finally, the body was wrapped in a heating pad. The chest was closed after the body temperature reached 30°C. At 35°C, the rat was taken off the anesthetic and housed in an oxygen chamber for 1 h. Healthy rats demonstrated an immediate recovery from the anesthetic. Efficiency and specificity of infection were assessed after 48–72 h.

2.1.2. Protocol 2. Gene transfer by partial isolation

We also tested the specificity and efficiency of gene delivery to the whole heart based on the cross-clamp method described by Hajjar *et al.* [4,18]. The heart is partially isolated *in vivo* by this approach, enabling a high intravascular pressure delivery of the adenovirus. In brief, rats were anesthetized as described above, and the chests were entered through a right thoracotomy. A catheter was inserted through the left ventricle and advanced to the aortic root. The aorta and pulmonary artery were cross-clamped distal to the catheter, and 0.2 ml viral solution (10^{12} viral particles/ml) was injected into the aortic root. The clamp was maintained for 30 s. After removing the clamp, the chest was closed and the animal recovered. Body temperature was maintained at 35–37°C during the procedure. Only those animals demonstrating the hallmark "blanching" of the heart during the viral injection were held for the analysis. Efficiency and specificity of infection were assessed after 48–72 h.

2.1.3. Assessing efficiency of gene transfer

At 48–72 h post gene transfer, efficiency was assessed after X-Gal staining of the whole heart for LacZ expression. Positive cells for LacZ expression turn blue with X-Gal staining. Efficiency was calculated as the percentage of blue cells.

The rats were heparinized and anesthetized with a lethal dose of pentobarbital. The hearts were excised, the aorta was canulated, and retrograde perfused with a series of syringes containing the following solutions:

(a) 20 ml PBS
(b) 20 ml 0.5% glutaraldehyde and 0.05% formaldehyde in PBS, then bathed in the same solution for 30 min, followed by a 20 ml PBS flush
(c) 20 ml 30% sucrose in PBS, 30 min bath in same solution, followed by a PBS flush
(d) 20 ml 0.01% sodium deoxycholate in PBS with 40 μl IGEPAL CA-630 detergent (SIGMA), 15 min bath in same solution, no final rinse
(e) 10 ml X-Gal (10 mg X-Gal (SIGMA) first dissolved in 0.5 ml dimethylformamide, then add 9.5 ml PBS). The heart was sliced into multiple cross sections and bathed in the X-Gal solution overnight. The stained cross sections are shown in Figs. 2 and 3.

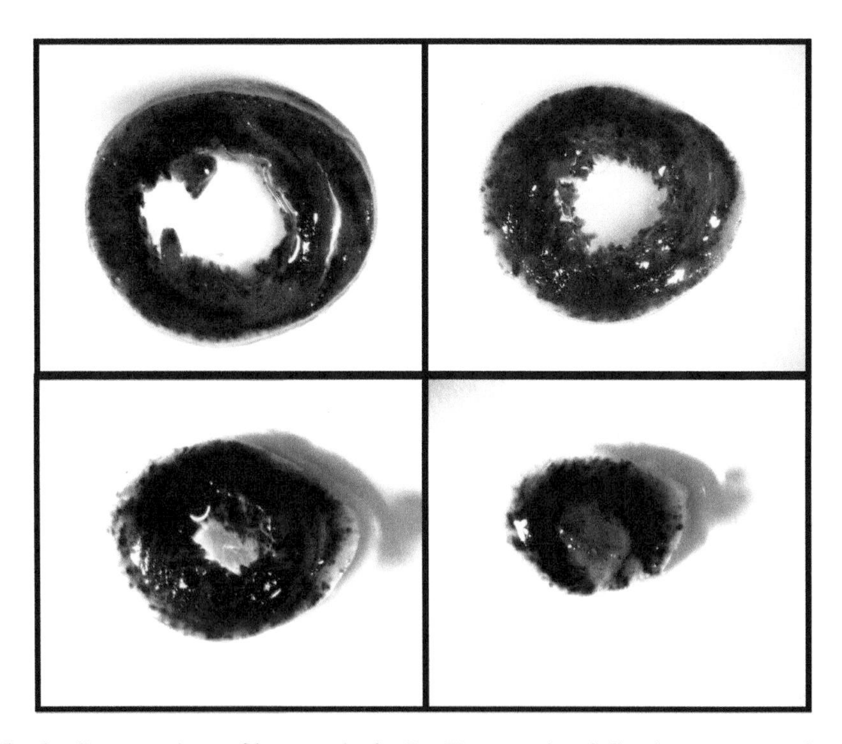

Fig. 2. Cross sections of heart stain for LacZ expression following gene transfer *in vivo* by Protocol 1; retrograde perfusion for 7.5 min with a calcium-free perfusate followed by a 90 s no flow viral exposure. Gene transfer efficiency is 60%.

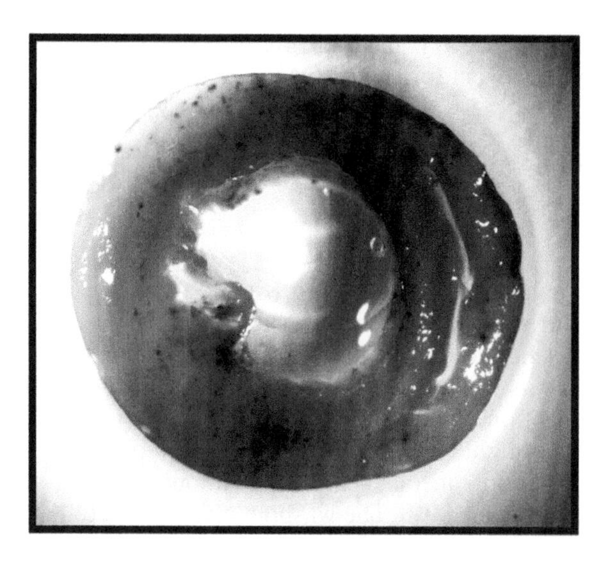

Fig. 3. Cross section of heart stained for LacZ expression following gene transfer *in vivo* by Protocol 2: 30 s aorta/pulmonary artery cross-clamp method. Gene transfer efficiency is $<5\%$.

The percentage of cells expressing the LacZ protein were quantified from those cross sections stained with X-Gal. Here, the cross sections were minced into 1 mm pieces and digested in collagenase (4 mg/ml dH_20, Worthington Biochemical Inc., Lakewood, NJ) for 6 h. The digestion was stirred throughout and triturated every 30 min. Undigested debris were filtered out and the remaining cells were collected by centrifugation, resuspended in PBS, and plated for observation by phase contrast microscope. Positive cells expressing the LacZ gene (blue cells) were counted along with total cells from multiple random views.

2.1.4. Assessing specificity of gene transfer

At 48–72 h post adenoviral delivery, LacZ expression was semi-quantitatively assessed in the left ventricle, liver, lung, and skeletal muscle based on a colorimetric assay for β-galactosidase activity (SIGMA kit GAL-A). Ad.cmv.LacZ-injected rats and PBS-injected sham-operated controls were heparinized, anesthetized with a lethal dose of pentobarbitol, and three 100 mg tissue sample from each organ were collected. Each sample was homogenized in 1 ml lysis buffer. The kit provided detailed instructions and all solutions required to complete the assay. There was one modification: blanks included an aliquot of the tissue homogenate (excluding the X-Gal reaction buffer). Finally, activity measurements were normalized to protein content. Protein content from the homogenate of each sample was determined by modification of the Lowry method (BCA protein assay reagent, Pierce Biotechnology Inc., Rockford, IL, USA).

3. RESULTS

3.1. Survival

As expected, the heart stopped beating while cross-clamped and retrograde perfused with the calcium-free Tyrode buffer according to Protocol 1. The heart was not ischemic during this 7.5 min perfusion period because the Tyrode solution was well oxygenated. Following this initial perfusion period and the viral injection, the hearts were perfused with Kreb's buffer containing normal levels of calcium. This step was required to reestablish calcium homeostasis and a spontaneous heartbeat. Without the washout phase of the protocol, mortality was 100%. The washout also provided the opportunity to flush unsequestered virus from the heart prior to releasing the cross-clamp.

Survival was dependent on the anesthetic. In initial experiments, the rats were anesthetized with sodium pentobarbitol. Recovery from anesthesia occurred over 3 h and survival was 75%. With isoflurane anesthesia, recovery was within 5–30 min, and survival was 90% for both Protocol 1 and 2. Importantly, the gas enabled control over the depth of the anesthetized state throughout the surgical procedure. Immediately after releasing the cross-clamp, the anesthetic was reduced from 1.5% to 0.5% isoflurane (v/v with 100% oxygen). This helped raise heart rate, contractility, and survival.

3.2. Efficiency

Adenoviral gene transfer was assessed by histochemistry (X-gal staining for LacZ expression) in the whole heart after 2–3 days of infection. As shown in Fig. 2, X-gal staining was homogeneously distributed throughout the multiple cross sections of the heart following gene delivery by Protocol 1. Figure 3 illustrates the level of gene transfer by Protocol 2. The expression was low and heterogeneous. Cells were isolated from stained hearts from both protocols to determine the efficiency of gene transfer, as the number of cells stained blue for exogenous gene expression relative to total cell counts. Efficiency was 58±11% for Protocol 1 and <5% for Protocol 2. The low level of gene transfer for Protocol 2 is consistent with earlier reports [3,10,11].

3.3. Specificity

The specificity of infection to the heart, compared to the liver, lung, and skeletal muscle, was assessed by an enzymatic assay for β-galactosidase activity. The results are shown in Fig. 4. In line with the histological analysis,

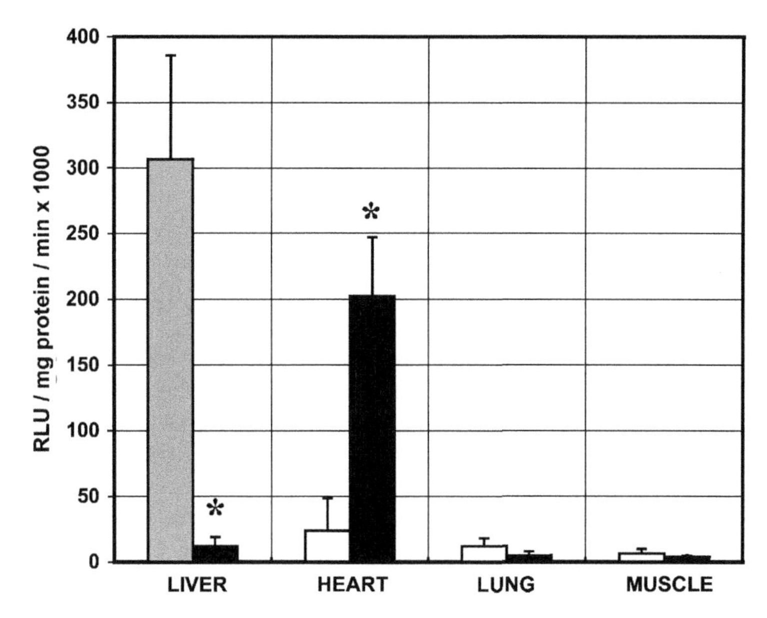

Fig. 4. Specificity of gene transfer: Protocol 1 (black bars) and Protocol 2 (grey bars). *β*-Gal activities in the liver, heart, lung, and skeletal muscle post Ad.cmv.LacZ infection. Protocol 1 (black bar) is the perfusion approach following complete isolation *in vivo*, and Protocol 2 (grey bar) is the aorta/pulmonary artery cross-clamp approach. Gene transfer was ten fold greater in hearts by Protocol 1 compared to Protocol 2. Infection of peripheral tissue by Protocol 1 was minimal because unsequestered virus was flushed from the heart prior to releasing the cross-clamp ($P<0.05$).

β-galactosidase enzymatic activity was ten fold greater in hearts infected by Protocol 1 compared to Protocol 2. In liver, *β*-Gal activity was dramatically lower by Protocol 1. The reduction in the infection of the liver is consistent with having flushed unsequestered virus out of the heart prior to releasing the cross-clamp.

4. DISCUSSION

In this study, we report a novel, catheter-based infection protocol that yields a highly efficient transfer of adenoviral genes throughout the rat heart *in vivo*. This infection approach is characterized by the following features: (i) the heart is entirely isolated *in vivo* by clamping all vessels to/from the heart simultaneously with a single clamp; (ii) a catheter positioned in the aortic root delivers the viral solution, while a second catheter positioned in the right ventricle provides a path for coronary efflux; (iii) the retrograde perfusion of the heart enables the blood to be flushed from the heart and replaced with well-oxygenated buffer containing the adenovirus and permeability agents; and (iv) at the end of the infection protocol, unsequestered virus is flushed from the heart prior to removing the clamp. Consequently, this approach results in a 58% infection of the whole heart *in vivo*, with minimal infection of the peripheral organs.

4.1. Efficiency – Gene transfer to the whole heart

There are a number of variables which affect the efficiency of gene transfer [14,19], including the concentration of virus, permeability enhancement strategy, perfusion pressure, temperature, blood vs. cardioplegic delivery solution, viral exposure times, viral serotype, and animal species. Optimizing many of these variables, Donahue showed that nearly 100% of the myocardium can undergo gene transfer in the isolated heart by retrograde perfusion with buffer containing the virus. The optimal conditions included a 30 min pretreatment period to enhance permeability with Krebs buffer containing serotonin and low calcium (50 mM), followed by a 2 min incubation period with the adenovirus at 2×10^9 pfu/ml [19]. While Donahue's work was done in the excised heart, we show that many of the same parameters can be controlled *in vivo* [9]. Because we are continuously perfusing the myocardium, we can control the viral concentration, permeability agent, cardioplegic solution, and temperature of the heart vs. body.

An important distinction between our approach and Donahue's is the infusion rate throughout the retrograde perfusion. Donahue found that the gene transfer to the whole heart increased with infusion rates up to 30 ml/min in the *ex vivo* model [14]. In preliminary studies, we found that the animals did not survive infusion rates at 30 ml/min. This result is consistent with injury data Emani *et al.* presented for infusion rates ranging between 1 and 30 ml/min [20]. They found that microvasculature injury was severe at 30 ml/min, and sparse at less than 5 ml/min. However, unlike Donahue, we are not delivering virus during this infusion rate. The retrograde perfusion is used here to enhance permeability, and a rate of <5 ml/min is sufficient to achieve this goal. We delivered the virus as a single bolus injection after the permeability period. The 0.4 ml viral cocktail was delivered within 2 s, generating a high intravascular pressure for gene transfer at 400 ± 100 mmHg. This adenoviral delivery strategy is similar to Hajjar's approach [4], and has been shown by several groups to have low mortality rates [3,9–11].

By this approach, we achieved 58% efficient gene transfer to the whole heart *in vivo*. This compares favorably to other gene transfer methods reported for rodent models, without the disadvantage of infecting peripheral tissue. Our approach represents a ten-fold increase over the aorta/pulmonary cross-clamp method (Protocol 2) for gene transfer in rat heart as described by Hajjar [4]. In hamster, Ikeda advanced Hajjar's approach by both treating the animals with a permeability agent (histamine) and increasing the viral incubation period to 5 min [3]. They achieved 77% efficient gene transfer to the whole heart *in vivo*. In mouse, Champion achieved 60% efficient gene transfer to the whole heart by clamping the descending aortic artery and maintaining an upper body perfusion circuit that contained the viral vector [12]. A less invasive approach in mouse has been demonstrated by injecting the virus into the left ventricle of the neonatal. By this closed chest procedure, Christensen *et al.* reported highly efficient and permanent gene expression [21]. A less invasive approach in adult rat has been reported by Ding *et al.* [11]. Here, the aortic root was accessed through the carotid artery in a closed chest model, and the efficiency of gene transfer to the whole heart was 43%. While the efficiency of gene transfer achieved by all these approaches in rodents are impressive, they all exhibited significant infection of peripheral tissue. Our approach overcomes this limitation by targeting the virus to the heart.

4.2. Specificity – Targeting gene transfer to the heart

The approach developed in our laboratory overcomes the limitation of the whole body infection. Because we are able to completely isolate the heart *in vivo*, we are able to infect and flush unsequestered virus from the heart prior to releasing the cross-clamp. As shown in Fig. 4, this dramatically reduces the infection of peripheral tissue to nearly undetectable levels.

Other catheter-based approaches of gene transfer fail to be cardiac specific because the viral vector remains in the systemic circulation following the viral delivery. As a potential solution, cardiac-specific promoters are being developed to target the gene expression to cardiomyocytes. However (i) the promoters are not 100% specific to the heart, (ii) gene expression is reduced in heart with the cardiac-specific promoter compared to the potent CMV promoter [22], and (iii) unsequestered virus still remains in the systemic circulation and accumulates in the liver and spleen regardless of the promoter [3,4,9,12].

4.3. Minimally invasive gene transfer – Promising technologies

While the adenovirus (rAd) can efficiently transduce the myocardium, the vector is limited by short-term gene expression (2 weeks) and an immune response to viral proteins [23], which can cause significant myocardial inflammation [8]. The recombinant adeno-associated viral vectors (rAAV) offer an alternative. The rAAV (25 nm) is smaller than the conventional rAd adenovirus allowing it to cross the endothelial barrier with greater ease [24,25]. Consequently, several groups have shown that simple intravenous or intramuscular injections yield highly efficient gene transfer to the heart or muscle [26–28]. This makes delivery both minimally invasive and highly efficient, two important criteria toward controlling gene transfer. In addition, the rAAV serotype is not pathogenic, there is no evidence of inflammation with use, and expression is long term (years) [29,30]. Unfortunately, as with the conventional adenovirus (rAd), expression of rAAV delivery genes within non-target cells remains problematic [30,31]. However, the various serotypes of the rAAV do have an array of tissue tropisms and binding characteristics [26,32], making the discovery of a cardiac-specific rAAV a promising possibility.

5. CONCLUSION

Much like the Wright brother's earliest achievements, whose designs were completely rethought and reworked, technical innovations continue to evolve in the field of gene transfer. Rapidly developing recombinant adenoviral vectors show great promise as an effective delivery vehicle. In this study, we report a new strategy for adenoviral gene delivery that can achieve highly efficient gene transfer to the myocardium with minimal infection of peripheral organs. This represents an important step toward achieving controlled gene transfer in the heart.

ACKNOWLEDGEMENTS

This study was supported by grants from the American Heart Association (O'Donnell; 0230099N) and the National Heart, Lung, and Blood Institute, NIH (O'Donnell; RO1 HL079415. Lewandowski; RO1HL 62702, RO1HL 56178, R37HL 49244). Figures reprinted with permission from the journal *Gene Therapy*.

REFERENCES

[1] E. C. Svensson, D. J. Marshall, K. Woodard, H. Lin, F. Jiang, L. Chu and J. M. Leiden, Efficient and stable transduction of cardiomyocytes after intramyocardial injection or intracoronary perfusion with recombinant adeno-associated virus vectors, *Circulation*, 1999, **99**, 201–205.

[2] A. S. Shah, R. E. Lilly, A. P. Kypson, O. Tai, J. A. Hata, A. Pippen, S. C. Silvestry, R. J. Lefk, D. D. Glower and W. J. Koch, Intracoronary adenovirus-mediated delivery and overexpression of the β-adrenergic receptor in the heart: prospects for molecular ventricular assistance, *Circulation*, 2000, **101**, 408–414.

[3] Y. Ikeda, Y. Gu, Y. Iwanaga, M. Hoshijima, S. S. Oh, F. J. Giordano, J. Chen, V. Nigro, K. L. Peterson, K. R. Chien and J. Ross, Restoration of deficient membrane proteins in the cardiomyopathic hamster by *in vivo* cardiac gene transfer, *Circulation*, 2002, **105**, 502–508.

[4] R. J. Hajjar, U. Schmidt, T. Matsui, J. L. Guerrero, K. H. Lee, J. K. Gwathmey, G. W. Dec, M. J. Semigran and A. Rosenzweig, Modulation of ventricular function through gene transfer *in vivo*, *Proc. Natl. Acad. Sci.*, 1998, **95**, 5251–5256.

[5] C. R. Bridges, K. Gopal, D. E. Holt, C. Yarnall, S. Cole, R. B. Anderson, X. Yin, A. Nelson, B. W. Kozyak, Z. Wang, J. Lesniewski, L. T. Su, D. M. Thesier, H. Sundar and J. H. Stedman, Efficient myocyte gene delivery with complete cardiac surgical isolation *in situ*, *J. Thorac. Cardiovasc. Surg.*, 2005, **130**(5), 1364.
[6] J. Fujishiro, H. Kawana, S. Inoue, H. Shimizu, H. Yoshino, Y. Hakamata, T. Kaneko, T. Murakami, K. Hashizume and E. Kobayashi, Efficiency of adenovirus-mediated gene transduction in heart grafts in rats, *Transplant Proc.*, 2005, **37**(1), 67–69.
[7] R. Bekeredjian, S. Chen, P. A. Frenkel, P. A. Grayburn and R. V. Shohet, Ultrasound-targeted microbubble destruction can repeatedly direct highly specific plasmid expression to the heart, *Circulation*, 2003, **108**, 1022–1026.
[8] E. Barr, J. Carroll, A. M. Kalynych, S. K. Tripathy, K. Kozarsky, J. M. Wilson and J. M. Leiden, Efficient catheter-mediated gene transfer into the heart using replication –defective adenovirus, *Gene Ther.*, 1994, **1**(1), 51–58.
[9] J. M. O'Donnell and E. D. Lewandowski, Efficient cardiac-specific adenoviral gene transfer in rat heart by isolated retrograde perfusion *in vivo*, *Gene Ther.*, 2005, **12**, 958–964.
[10] M. J. Wright, L. M. L. Wightman, D. S. Latchman and M. S. Marber, *In vivo* myocardial gene transfer: optimization and evaluation of intracoronary gene delivery *in vivo*, *Gene Ther.*, 2001, **8**, 1833–1839.
[11] Z. Ding, C. Cach, A. Sasse, A. Godecke and J. Schrader, A minimally invasive approach for efficient gene delivery to rodent hearts, *Gene Ther.*, 2004, **11**, 260–265.
[12] H. C. Champion, D. Georgakopoulos, S. Haldar, L. Wang, Y. Wang and D. A. Kass, Robust adenoviral and adeno–associated viral gene transfer to the *in vivo* murine heart, *Circulation*, 2003, **108**, 2790–2797.
[13] M. Iwatate, Y. Gu, T. Dieterle, Y. Iwanaga, K. L. Peterson, M. Hoshijima, K. R. Chien and J. Ross, *In vivo* high-efficiency transcoronary gene delivery and Cre-LoxP gene switching in the adult mouse heart, *Gene Ther.*, 2003, **10**(21), 1814–1820.
[14] J. K. Donahue, K. Kikkawa, D. C. Johns, E. Marban and J. H. Lawrence, Ultrarapid, highly efficient viral gene transfer to the heart, *Proc. Natl. Acad. Sci.*, 1997, **94**, 4664–4669.
[15] T. C. Becker, R. J. Noel, W. S. Coats, A. M. Gomez-Foix, T. Alam, R. D. Gerard and C. B. Newgard, Use of recombinant adenovirus for metabolic engineering of mammalian cells, *Methods Cell Biol.*, 1994, **43**(Pt A), 161–189 (Review).
[16] T. C. He, S. Zhou, L. T. Da Costa, J. Yu, K. W. Kinzler and B. Vogelstein, A simplified system for generating recombinant adenoviruses, *Proc. Natl. Acad. Sci.*, 1998, **95**, 2509–2514.
[17] D. Logeart, S. N. Hatem, C. Rucker-Martin, N. Chossat, N. Nevo, H. Haddada, M. Heimburger, M. Perricaudet and J. J. Mercadier, Highly efficient adenovirus-mediated gene transfer to cardiac myocytes after single-pass coronary delivery, *Hum. Gene Ther.*, 2000, **11**, 1015–1022.
[18] F. Del Monte and R. J. Hajjar, in *Efficient viral gene transfer to rodent hearts in vivo* (eds. J. M. Metzger), Methods in Molecular Biology, Humana Press Inc, Totowa, NJ, 2003, Vol. 219, pp. 179–193.
[19] J. K. Donahue, K. Kikkawa, A. D. Thomas, E. Marban and J. H. Lawrence, Acceleration of widespread adenoviral gene transfer to intact rabbit hearts by coronary perfusion with low calcium and serotonin, *Gene Ther.*, 1998, **5**, 630–634.
[20] S. M. Emani, A. S. Shah, M. K. Bowman, S. Emani, K. Wilson, D. D. Glower and W. J. Koch, Catheter-based intracoronary myocardial adenoviral gene delivery: importance of intraluminal seal and infusion flow rate, *Mol. Ther.*, 2003, **8**(2), 306–313.
[21] G. Christensen, S. Minamisawa, P. J. Gruber, Y. Wang and K. R. Chien, High-efficiency, long-term cardiac expression of foreign genes in living mouse embryos and neonates, *Circulation*, 2000, **101**(2), 178–184.

[22] H. Ma, C. M. Sumbilla, I. K. Farrance, M. G. Klein and G. Inesi, Cell-specific expression of SERCA, the exogenous Ca^{2+} transport ATPase, in cardiac myocytes, *Am. J. Physiol. Cell Physiol.*, 2004, **286**(3), C556–C564.
[23] H. Gilgenkrantz, D. Duboc, V. Juillard, D. Couton, A. Pavirani, J. G. Guillet, P. Briand and A. Kahn, Transient expression of genes transferred *in vivo* into heart using first-generation adenoviral vectors: role of the immune response, *Hum. Gene Ther.*, 1995, **6**(10), 1265–1274.
[24] Q. Xie, W. Bu, S. Bhatia, J. Hare, T. Somasundaram, A. Azzi and M. S. Chapman, The atomic structure of adeno-associated virus (AAV-2), a vector for human gene therapy, *Proc. Natl. Acad. Sci.*, 2002, **99**, 10405–10410.
[25] D. Chu, C. C. Sullivan, M. D. Weitzman, L. Du, P. L. Wolf, S. W. Jamieson and P. A. Thistlethwaite, Direct comparison of efficiency and stability of gene transfer into the mammalian heart using adeno-associated virus versus adenovirus vectors, *J. Thorac. Cardiovasc. Surg.*, 2003, **126**(3), 671–679.
[26] T. Zhu, L. Zhou, S. Mori, Z. Wang, C. F. McTiernan, C. Qiao, C. Chen, D. W. Wang, J. Li and X. Xiao, Sustained whole-body functional rescue in congestive heart failure and muscular dystrophy hamsters by systemic gene transfer, *Circulation*, 2005, **112**(17), 2650–2690.
[27] M. J. Blankinship, P. Gregorevic, J. M. Allen, S. Q. Harper, H. Harper, C. L. Halber, A. D. Miller and J. S. Chamberlain, Efficient transduction of skeletal muscle using vectors based on adeno-associated virus serotype 6, *Mol. Ther.*, 2004, **10**(4), 671–678.
[28] S. Kawamoto, Q. Shi, Y. Nitta, J. Miyazaki and M. D. Allen, Widespread and early myocardial gene expression by adeno-associated virus vector type 6 with a beta-actin hybrid promoter, *Mol. Ther.*, 2005, **11**(6), 980–985.
[29] P. Gregorevic, M. J. Blankinship and J. S. Chamberlain, Viral vectors for gene transfer to striated muscle, *Curr. Opin. Mol. Ther.*, 2004, **6**(5), 491–498 (Review).
[30] R. Aikawa, G. S. Huggins and R. O. Snyder, Cardiomyocyte-specific gene expression following recombinant adeno-associated viral vector transduction, *J. Biol. Chem.*, 2002, **21**, 18979–18985.
[31] H. Su, S. Joho, Y. Huang, A. Barcena, J. Arakawa-Hoyt, W. Grossman and Y. W. Kan, Adeno-associated viral vector delivers cardiac-specific and hypoxia-inducible VEGF expression in ischemic mouse hearts, *Proc. Natl. Acad. Sci.*, 2004, **101**(46), 16280–16285.
[32] L. Du, M. Kido, D. V. Lee, J. E. Rabinowitz, R. J. Samulski, S. W. Jamieson, M. D. Weitzman and P. A. Thistlethwaite, Differential myocardial gene delivery by recombinant serotype-specific adeno-associated viral vectors, *Mol. Ther.*, 2004, **10**(3), 604–608.

Biological and Cellular Barriers Limiting the Clinical Application of Nonviral Gene Delivery Systems*

Régis Cartier[1] and Regina Reszka[2]

[1]*Nanotechnology Group, Clinic of Anaesthesiology and Operative Medicine Charite, Berlin University Medical School, Berlin, Germany*
[2]*Helios Research Center, Wiltbergstrasse 50, 13125 Berlin-Buch, Germany*

Abstract

The clinical success of nonviral gene therapy will depend on our ability to design efficient and targeted gene delivery systems. It is now understood that the gene transfer mechanism is a complicated process during which the vector has to overcome numerous biological and cellular barriers. As a consequence, nonviral gene delivery systems tend to multicomponent complexes containing various biofunctional elements. The ultimate goal is to circumvent inhibitory effects and to take advantage of cellular transport and degradation processes. The challenges for the development of novel systems are multiple. A better understanding of the molecular processes underlying the transfection mechanism is required. It is not clear through which mechanism the vector enters the cell and finally reaches the nucleus. In addition, cellular factors limiting the transfection process should also be identified. Because of the critical biological role of DNA, the existence of specific protective cellular mechanisms, as the result of the massive uptake of exogenous DNA, is expected. Another important issue is the influence of physicochemical properties of DNA complexes such as size, condensation rate and charge on the transfection process. These parameters are of contradictory effects depending on the particular gene transfer step. Optimization of the transfection system thus comprises chemical modification and/or incorporation of functional elements as well as control of the physicochemical parameters of the complex during the transfection process. The controlled assembly of a multicomponent system and the proper unfolding of its functional elements present a considerable challenge. Their success will require a close cooperation between biologists and material scientists.

Keywords: Nonviral gene transfer, clinical application, gene transfer mechanism, peptides, heteroplex, artificial virus-like particles.

Contents

1. Introduction 48
2. The gene transfer mechanism 48
3. Specificity of targeting peptides during transfection 50
4. The need of standardization 51
5. The effect of physicochemical properties of DNA complexes on transfection 52
6. Understanding the behavior of DNA complexes inside the cell 53

---

*This paper was first published in Gene Therapy, 2002 February, 9(3), 157–167.

7. Rational design of heteroplexes with *in vivo* application 54
References 55

1. INTRODUCTION

The goal of gene therapy is either the replacement of a defective human gene or the introduction of a new gene encoding a therapeutic protein. To be successful, this requires both delivery of the therapeutic DNA into human cells and its subsequent expression. It is widely recognized that the transfer of exogenous DNA into somatic cells is a complex process involving the passage through numerous biological barriers, and is particularly complicated in the case of *in vivo* applications. These barriers include the extracellular matrix, the cell membrane, the intracellular environment and also the nuclear envelope.

To overcome these barriers, the development of nonviral gene transfer systems tends toward multicomponent structures which may contain (poly)cationic elements for DNA condensation, lipidic compounds for enhanced protection and affinity to cell membranes and additional ligands for cell targeting and intracellular processing. Because of their structural and functional properties they are described as artificial virus-like particles and will be termed herein as heteroplexes in analogy to lipo- and polyplexes [1]. It is claimed that heteroplexes can circumvent the current limiting factors of viral vectors such as immunogenicity, size limitations of the transgene, potential mutation to replication-competent recombinant viruses and the expensive production. However, nonviral gene transfer systems still suffer from a relatively low transfection efficiency compared with viral vectors, hindering their broad clinical application.

2. THE GENE TRANSFER MECHANISM

Following an *in vivo* application the vector encounters different barriers depending on the particular route of administration. In the case of systemic injection, interaction with plasma proteins can take place leading to aggregation of the vector and subsequently, to accumulation in the liver or the lung before the vector reaches its primary target. An additional limiting factor is the clearance from the blood by the reticuloendothelial system. Several efforts were focused on the chemical modification of the heteroplex surface in order to generate long-circulating particles. One widely used approach is the use of polyethyleneglycol to build a highly hydrophilic coat, which stabilizes the particle by steric-repulsion forces. For targeting of the circulating complex to the site of delivery, an additional targeting moiety is attached on the particle surface. When the vector is injected directly into the targeted organ, transfection efficiency can be enhanced using physical methods such as *in vivo* electroporation [2,3], increase of hydrodynamic pressure [4,5], particle gun [6,7] and jet injection [8]. The latter has recently been applied to tumors *in vivo* [9].

The transfection mechanism of the targeted cell can be described as a sequence of several transport and dismantling steps of the heteroplex and are depicted in Table 1.

The initial step is the binding of the heteroplex to the cell surface and its subsequent internalization. For this purpose, ligands or antibodies specific to cell membrane receptors can be displayed on the heteroplex surface to enhance cell type-specific targeting and cell entry *via* receptor-mediated endocytosis. However, one cannot exclude the possibility of direct entry into the cell by transient permeabilization of the cell membrane, especially when heteroplexes containing liposomes or when amphiphilic peptides are used [10,11]. The next step

Table 1. Left, schematic representation at the cellular level. Factors limiting the transfection efficiency (middle) and mechanisms, which could potentially enhance targeting and efficiency (right) of the heteroplex are indicated

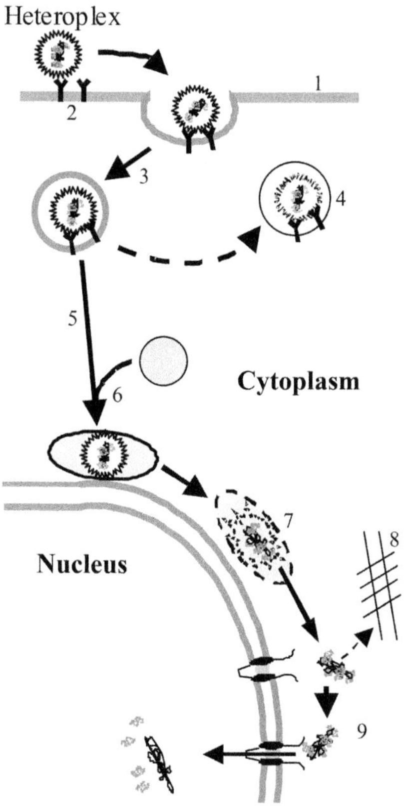

Gene transfer mechanism	Limiting factors	Possible enhancing factors
	Cell membrane (1)	Cell surface receptors (2)
		Endocytosis (3)
	Recycling endosomes (4)	
		Vesicular transport (5)
	Fusion with lysosomes and degradation (6)	
	Cytoplasmic entrapment (8)	pH-driven escape from the lysosome (7)
		Nuclear import machinery (9)

concerns the intracellular transport of the heteroplex up to the nucleus. The endocytotic pathway potentially offers vesicular transport to the perinuclear region. However, this requires proper sorting of the endosome, a process dependent on physicochemical properties of the endocytosed material [12]. Different strategies are used to allow the release of the vector into the cytoplasm, including the application of endosomolytic peptides or lipids [11,13] and polyethyleneimines acting as a protonic sponge and inducing osmotic disruption of the endosome [14,15]. Once in the cytoplasm, the heteroplex is exposed to soluble and insoluble factors, which may lead to cytoplasmic entrapment and subsequent degradation. It is also expected that intracellular transport and degradation processes which interfere with the gene transfer mechanism will depend on the cell physiology and therefore also on the specific cell type and its microenvironment. The last step is finally the transport into the nucleus and the release of the DNA [16], in order to permit transcription of the therapeutic gene. One approach is to avoid nuclear import altogether by developing a cytoplasmic expression system based on the T7 RNA polymerase [17,18]. Another strategy is to take advantage of the nucleocytoplasmic transport machinery. The general proof of this concept is provided by the fact that some viruses, among them DNA viruses, as part of their life cycle transport their genome through the nuclear pore complex (NPC), probably by interaction with the cellular nuclear import machinery [19,20]. However, this crucial translocation step of the viral DNA is not completely understood so that it cannot be transposed directly to nonviral gene transfer systems. Nevertheless, numerous efforts were made to modify plasmid DNA so that it can be recognized by cellular factors as a nuclear import substrate. Modifications include additional cloning of specific DNA sequences recognized by transcription factors [21], the attachment of glycosylated moieties [22] and synthetic peptides acting as nuclear localization signals (NLS). Over the past few years, tremendous advances have been realized elucidating the nuclear transport mechanism of macromolecules. It resulted in the identification of a wide range of potential NLS sequences and possibilities for regulation, which may be applied to drug delivery systems [23].

3. SPECIFICITY OF TARGETING PEPTIDES DURING TRANSFECTION

A major challenge in the utilization of targeting molecules such as peptides is to avoid unspecific effects during the transfection process. In numerous studies, NLS sequences have been incorporated into transfection systems with the rationale to facilitate the transport of DNA from the cytoplasm into the nucleus *via* the nuclear import machinery in an NLS sequence-specific manner. Not all studies included experiments using an NLS-control peptide to demonstrate sequence specificity. In these cases, one cannot exclude the possibility that the incorporated peptide may enhance the transfection efficiency through a mechanism other than the nuclear import machinery. It is likely that the basic residues of the NLS interact directly with the DNA contributing to its condensation and increasing the total charge of the particular transfection system. Recent studies suggest that cell binding and internalization of DNA complexes is mediated by heparan sulfate proteoglycans (HSPGs) [24–26]. These molecules are displayed on the cell surface and bind to positively charged DNA complexes through nonspecific electrostatic interactions. The expression of HSPG depends on many factors including the cell type and cell differentiation. HSPG is also found intracellularly and especially in the nucleus [27]. Furthermore, proteoglycans harboring different types of glycosaminoglycans (GAGs) are major components of the extracellular matrix in many organs. Interaction of lipoplexes and polyplexes with GAGs was demonstrated *in vitro* [28] and may lead to entrapment or dissociation of the DNA complex inside the organ. Coating of the polyplex with an anionic lipid prevented the inhibition of polyfection *in vitro* by hyaluronic acid [29]. Taken together, it is possible that NLS-containing peptides incorporated in heteroplexes interact with proteoglycans. This interaction may enhance the cell entry of the heteroplex and possibly interferes with other steps of the gene transfer mechanism. The anionic residues of an NLS-containing peptide could also interact with the polycations typically used for DNA condensation or the hydrophobic residues with lipid components. The NLS could thereby influence the physical properties of the heteroplex such as particle size, DNA condensation rate, surface-charge distribution and intracellular stability. These effects could be sequence specific and could influence the transfection efficiency of the complex. Therefore, it is important to design studies that permit the detection of nuclear import selectively. So far, only four different NLS sequences have been used in combination with gene transfer systems and most of the analyzed studies involve the well characterized SV40 T large antigen NLS-sequence ^{126}PKKKRKV^{132}. As a result of its high content of lysine and arginine residues, it increases the cationic residues: DNA ratio within the complex. It will be important to include NLS sequences with different physicochemical properties, e.g., bipartite NLS consisting of two basic clusters separated by a linker region. The bipartite NLS binds importin receptor at multiple sites [30], offering greater sequence diversity [31]. Furthermore, a broad variety of other types of NLS sequences have been identified with different corresponding receptors and potentially different nuclear import pathways that may be of interest for gene transfer systems.

4. THE NEED OF STANDARDIZATION

Over the past decade, a tremendous number of studies on the development of multifunctional gene delivery systems were published. Most of the studies differ widely in several critical aspects including the used cell types and transfection systems, and also in the experimental model systems and approaches employed to investigate the role of the functional elements during transfection. A detailed analysis and meaningful comparison of the results obtained in order to derive clear conclusions describing the current state of the art, remain thus a difficult task.

Therefore, certain standardization or a detailed description of factors known to influence transfection efficiency is believed to be of great value for future studies. These may include at least the following points:

(1) analysis and assessment of parameters describing the purity of DNA;
(2) storage and handling of synthetic peptides;
(3) detailed procedure of DNA complex formation;
(4) methods for the physicochemical characterization of DNA complexes and their validation; and
(5) general definition of what should be valuable control experiments, particularly regarding peptide sequence specificity.

In addition to these specific considerations, one may also emphasize the following general points, which are particularly critical for the utilization of gene transfer technologies:

(1) Standard cell culture conditions for a given cell type or definition of experimental criteria in order to describe physiological parameters such as growth rate and confluency. Information on the origin and the number of passages of the considered cell line, as well as frequencies of mycoplasma decontamination may also be of interest.
(2) Accurate statistical analysis of experimental data. Precise information about the statistical method and the corresponding preliminary validation test would be helpful if presented in a clear manner.

5. THE EFFECT OF PHYSICOCHEMICAL PROPERTIES OF DNA COMPLEXES ON TRANSFECTION

The gene transfer process is largely influenced by the chemical composition of the heteroplex. However, physicochemical properties also determine the fate of the gene transfer system. These properties include size, surface charge and DNA condensation rate of the complex and must be optimized to overcome a specific biological barrier. A major difficulty in this optimization process is the observation that physicochemical properties have contradictory effects depending on the particular gene transfer step. A high DNA condensation rate is required for efficient cellular uptake through endocytosis and protection from enzymatic degradation. However, for efficient expression the DNA molecule must be decondensated to permit recognition by the transcription machinery.

The role of size and conformation state of the DNA for nuclear uptake is not fully understood. Transport into the nucleus, *via* the NPC, occurs for a broad variety of macromolecules including proteins and nucleic acids. For both, size dependence seems to be a common theme. Indeed, passive diffusion through the NPC is observed for proteins smaller than 60 kDa and oligodinucleotides (ODN) or small DNA fragments up to 200 bp [32,33]. With increasing size, diffusion becomes weaker, but nuclear accumulation can be actively enhanced by taking advantage of the NLS-dependent nuclear import machinery. This means that the cell's machinery actually accelerates the transport process of molecules, which can be classified as 'permissive' for nuclear uptake. Extensive microinjection studies using linear DNA fragments showed that size is a major factor limiting cytoplasmic mobility [34] and transport into the nucleus [32], which may lead to cytoplasmic entrapment and subsequently to degradation. However, it remains unclear whether size is also a significant factor limiting intracellular transport processes of plasmid DNA, and particularly of DNA complexes used in transfection systems. Circular and supercoiled DNA is believed to be the predominant form of a transfection-active plasmid DNA. Their hydrodynamic diameter is therefore highly reduced as compared with linear DNA, especially when additional condensing agents are used. Furthermore, endocytosed DNA complexes can be localized in the perinuclear region after a few hours following transfection [35,36], suggesting that a high number of plasmid DNA molecules are closely localized to the nuclear envelope. Finally, molecules commonly used for transfection are smaller than 10 kb. In comparison, a DNA virus such as the herpes simplex virus 1 is able to transport its 152-kb genome through the NPC [20]. One may therefore hypothesize that additional factors beyond a simple size constraint hinder nuclear accumulation of plasmid DNA. Plasmid DNA may belong to a class of 'nonpermissive' molecules for nuclear uptake, against which the cell has developed mechanisms for exclusion from the nucleus. Potential mechanisms may include cytoplasmic retention, down-regulation of NPC permeability and rapid nuclear export.

6. UNDERSTANDING THE BEHAVIOR OF DNA COMPLEXES INSIDE THE CELL

To date, little is known about the transport process of plasmid DNA or DNA complexes into the nucleus. Since large aggregates are supposedly excluded from the nucleus, single DNA molecules or perhaps small DNA complexes must be released at some point during transfection. It is unknown whether plasmid DNA enters the nucleus in a circular and condensed form or whether cellular factors are involved in the translocation process. Cellular factors have to be identified that physically interact with

DNA and thereby hinder nuclear transfer either directly or by inducing additional cellular mechanisms. Based on this knowledge, specific modifications of the DNA could produce molecules 'permissive' for nuclear localization and with an increased translocation rate through the NPC. The current belief is that DNA complexes utilize the endocytotic pathway for cell entry and intracellular traffic and subsequently gain entry into the nucleus during mitosis. However, the sensitivity of most experimental methods only allows tracking of relatively large DNA complexes. Therefore, one cannot exclude the existence of other pathways, which may effectively lead to transfection. A recent study demonstrates that although microinjected plasmid DNA accumulated in the nucleus primarily in cells which underwent mitosis, a number of DNA molecules were able to translocate into the nucleus in nonmitotic cells [37].

7. RATIONAL DESIGN OF HETEROPLEXES WITH *IN VIVO* APPLICATION

Synthetic elements derived from the elucidation of the gene transfer mechanism must be assembled in an appropriate way so that each component retains its expected activity within the heteroplex. Furthermore, the physicochemical properties of heteroplexes also depend on the production procedure and determine the intracellular fate. Hence, the ability to build heteroplexes with defined physicochemical properties is a critical issue in the design of multicomponent transfection systems. Another important aspect concerns the applicability of the heteroplex *in vivo*. As mentioned earlier, the heteroplex is subjected to numerous extracellular factors depending on the administration route. In the case of intravenous delivery, the heteroplex may interact with blood elements including albumin, opsonins and the complement system, altering its resistance to degradation. A previous study based on 2D-PAGE analysis demonstrated that interaction of liposomes with plasma proteins could be altered upon specific surface modifications and also influenced subsequent organ distribution *in vivo* [38]. Therefore, the cell entry mechanism and intracellular fate may vary significantly to that expected from cell culture experiments. Consequently, optimization of a heteroplex only makes sense if the cell entry mechanism *in vivo* and the properties of the internalized material are well characterized and reproducible in the *in vitro* model systems or in the cell culture.

To conclude, the successful design of novel heteroplexes suitable for an *in vivo* gene therapy will require a better understanding of both extra- and intracellular processes and synergistic efforts with polyelectrolyte chemists and colloid scientists.

REFERENCES

[1] P. L. Felgner, Y. Barenholz, J. P. Behr, S. H. Cheng, P. Cullis, L. Huang, J. A. Jessee, L. Seymour, F. Szoka, A. R. Thierry, E. Wagner and G. Wu, Nomenclature for synthetic gene delivery systems, *Hum. Gene Ther.*, 1997, **8**, 511–512.

[2] T. Nishi, K. Yoshizato, S. Yamashiro, H. Takeshima, K. Sato, K. Hamada, L. Kitamura, T. Yoshimura, H. Saya, J. Kuratsu and Y. Ushio, High-efficiency *in vivo* gene transfer using intraarterial plasmid DNA injection following *in vivo* electroporation, *Cancer Res.*, 1996, **56**, 1050–1055.

[3] H. Aihara and J. Miyazaki, Gene transfer into muscle by electroporation *in vivo*, *Nat. Biotechnol.*, 1998, **16**, 867–870.

[4] G. Zhang, D. Vargo, V. Budker, N. Armstrong, S. Knechtle and J. A. Wolff, Expression of naked plasmid DNA injected into the afferent and efferent vessels of rodent and dog livers, *Hum. Gene Ther.*, 1997, **8**, 1763–1772.

[5] F. Liu, Y. Song and D. Liu, Hydrodynamics-based transfection in animals by systemic administration of plasmid DNA, *Gene Ther.*, 1999, **6**, 1258–1266.

[6] W. H. Sun, J. K. Burkholder, J. Sun, J. Culp, J. Turner, X. G. Lu, T. D. Pugh, W. B. Ershler and N. S. Yang, *In vivo* cytokine gene transfer by gene gun reduces tumor growth in mice, *Proc. Natl. Acad. Sci. USA*, 1995, **92**, 2889–2893.

[7] A. L. Rakhmilevich, K. Janssen, J. Turner, J. Culp and N. S. Yang, Cytokine gene therapy of cancer using gene gun technology: Superior antitumor activity of interleukin-12, *Hum. Gene Ther.*, 1997, **8**, 1303–1311.

[8] P. A. Furth, D. Kerr and R. Wall, Gene transfer by jet injection into differentiated tissues of living animals and in organ culture, *Mol. Biotechnol.*, 1995, **4**, 121–127.

[9] R. Cartier, S. V. Ren, W. Walther, U. Stein, A. Lewis, P. M. Schlag, M. Li and P. A. Furth, *In vivo* gene transfer by low-volume jet injection, *Anal. Biochem.*, 2000, **282**, 262–265.

[10] C. W. Pouton and L. W. Seymour, Key issues in non-viral gene delivery, *Adv. Drug Deliv. Rev.*, 1998, **34**, 3–19.

[11] C. Plank, W. Zauner and E. Wagner, Application of membrane-active peptides for drug and gene delivery across cellular membranes, *Adv. Drug Deliv. Rev.*, 1998, **134**, 21–35.

[12] C. M. Varga, T. J. Wickham and D. A. Lauffenburger, Receptor-mediated targeting of gene delivery vectors: Insights from molecular mechanisms for improved vehicle design, *Biotechnol. Bioeng.*, 2000, **70**, 593–605.

[13] J. T. Sparrow, V. V. Edwards, C. Tung, M. J. Logan, M. S. Wadhwa, J. Duguid and L. C. Smith, Synthetic peptide-based DNA complexes for nonviral gene delivery, *Adv. Drug Deliv. Rev.*, 1998, **30**, 115–131.

[14] O. Boussif, F. Lezoualc'h, M. A. Zanta, M. D. Mergny, D. Scherman, B. Demeneix and J. P. Behr, A versatile vector for gene and oligonucleotide transfer into cells in culture and *in vivo*: Polyethylenimine, *Proc. Natl. Acad. Sci. USA*, 1995, **92**, 7297–7301.

[15] A. Kichler, C. Leborgne, E. Coeytaux and O. Danos, Polyethylenimine-mediated gene delivery: A mechanistic study, *J. Gene Med.*, 2001, **2**, 135–144.

[16] D. V. Schaffer, N. A. Fidelman, N. Dan and D. A. Lauffenburger, Vector unpacking as a potential barrier for receptor-mediated polyplex gene delivery, *Biotechnol. Bioeng.*, 2000, **67**, 598–606.

[17] X. Gao and L. Huang, Cytoplasmic expression of a reporter gene by co-delivery of T7 RNA polymerase and T7 promoter sequence with cationic liposomes, *Nucleic Acids Res.*, 1993, **21**, 2867–2872.

[18] M. Brisson, W. C. Tseng, C. Almonte, S. Watkins and L. Huang, Subcellular trafficking of the cytoplasmic expression system, *Hum. Gene Ther.*, 1999, **10**, 2601–2613.

[19] H. Kasamatsu and A. Nakanishi, How do animal DNA viruses get to the nucleus? *Annu. Rev. Microbiol.*, 1998, **52**, 627–686.

[20] G. R. Whittaker, M. Kann and A. Helenius, Viral entry into the nucleus, *Annu. Rev. Cell Dev. Biol.*, 2000, **16**, 627–651.
[21] J. Vacik, B. S. Dean, W. E. Zimmer and D. A. Dean, Cell-specific nuclear import of plasmid DNA, *Gene Ther.*, 1999, **6**, 1006–1014.
[22] I. Fajac, P. Briand, M. Monsigny and P. Midoux, Sugar-mediated uptake of glycosylated polylysines and gene transfer into normal and cystic fibrosis airway epithelial cells, *Hum. Gene Ther.*, 1999, **10**, 395–406.
[23] D. A. Jans, C. K. Chan and S. Huebner, Signals mediating nuclear targeting and their regulation: Application in drug delivery, *Med. Res. Rev.*, 1998, **18**, 189–223.
[24] K. A. Mislick and J. D. Baldeschwieler, Evidence for the role of proteoglycans in cation-mediated gene transfer, *Proc. Natl. Acad. Sci. USA*, 1996, **93**, 12349–12354.
[25] L. C. Mounkes, W. Zhong, G. Cipres-Palacin, T. D. Heath and R. J. Debs, Proteoglycans mediate cationic liposome-DNA complex-based gene delivery *in vitro* and *in vivo*, *J. Biol. Chem.*, 1998, **273**, 26164–26170.
[26] C. M. Wiethoff, J. G. Smith, G. S. Koe and C. R. Middaugh, The potential role of proteoglycans in cationic lipid-mediated gene delivery. Studies of the interaction of cationic lipid–DNA complexes with model glycosaminoglycans, *J. Biol. Chem.*, 2001, **276**, 32806–32813.
[27] T. P. Richardson, V. Trinkaus-Randall and M. A. Nugent, Regulation of heparan sulfate proteoglycan nuclear localization by fibronectin, *J. Cell Sci.*, 2001, **114**, 1613–1623.
[28] M. Ruponen, S. Yla-Herttuala and A. Urtti, Interactions of polymeric and liposomal gene delivery systems with extracellular glycosaminoglycans: Physicochemical and transfection studies, *Biochim. Biophys. Acta*, 1999, **1415**, 331–341.
[29] E. Mastrobattista, R. H. Kapel, M. H. Eggenhuisen, P. J. Roholl, D. J. Crommelin, W. E. Hennink and G. Storm, Lipid-coated polyplexes for targeted gene delivery to ovarian carcinoma cells, *Cancer Gene Ther.*, 2001, **8**, 405–413.
[30] M. R. Fontes, T. The and B. Kobe, Structural basis of recognition of monopartite and bipartite nuclear localization sequences by mammalian importin-alpha, *J. Mol. Biol.*, 2000, **297**, 1183–1194.
[31] M. R. Hodel, A. H. Corbett and A. E. Hodel, Dissection of a nuclear localization signal, *J. Biol. Chem.*, 2001, **276**, 1317–1325.
[32] J. J. Ludtke, G. F. Zhang, M. G. Sebestyen and J. A. Wolff, A nuclear localization signal can enhance both the nuclear transport and expression of 1 kb DNA, *J. Cell Sci.*, 1999, **112**, 2033–2041.
[33] J. P. Leonetti, N. Mechti, G. Degols, C. Gagnor and B. Lebleu, Intracellular distribution of microinjected antisense oligonucleotides, *Proc. Natl. Acad. Sci. USA*, 1991, **88**, 2702–2706.
[34] G. L. Lukacs, P. Haggie, O. Seksek, D. Lechardeur, N. Freedman and A. S. Verkman, Size-dependent DNA mobility in cytoplasm and nucleus, *J. Biol. Chem.*, 2000, **275**, 1625–1629.
[35] V. Escriou, C. Ciolina, A. Helbling-Leclerc, P. Wils and D. Scherman, Cationic lipid-mediated gene transfer: Analysis of cellular uptake and nuclear import of plasmid DNA, *Cell Biol. Toxicol.*, 1998, **14**, 95–104.
[36] A. Coonrod, F. Q. Li and M. Horwitz, On the mechanism of DNA transfection: Efficient gene transfer without viruses, *Gene Ther.*, 1997, **4**, 1313–1321.
[37] V. Escriou, M. Carriere, F. Bussone, P. Wils and D. Scherman, Critical assessment of the nuclear import of plasmid during cationic lipid-mediated gene transfer, *J. Gene Med.*, 2001, **3**, 179–187.
[38] W. E. Bucke, S. Leitzke, J. E. Diederichs, K. Borner, H. Hahn, S. Ehlers and R. H. Muller, Surface-modified amikacin-liposomes: Organ distribution and interaction with plasma proteins, *J. Drug Target*, 1998, **5**, 99–108.

Designing Polymer-Based DNA Carriers for Non-Viral Gene Delivery: Have We Reached an Upper Performance Limit?

Jean-François Lutz

Research Group Nanotechnology for Life Science, Fraunhofer Institute for Applied Polymer Research, Geiselbergstrasse 69, Golm 14476, Germany

Abstract

The present chapter reports the point of view of a polymer scientist in the field of non-viral gene delivery. Hence, the material science aspects of this therapeutic field are primarily discussed herein. In particular, this chapter reviews recent progress for building high-performance polymer-based gene carriers. The fundamental relation between the structure of these polymeric vectors and their performances *in vitro* or *in vivo* is presented. This state-of-the-art review illustrates that the strong limitations of polymeric gene carriers can be potentially overcome if a careful design is made at the molecular scale or at the nanoscale. Thus, this chapter serves as a base for further discussions about improvements in non-viral gene delivery.

Keywords: Non-viral gene delivery, polycations, DNA, polyplexes, synthetic polymers.

Contents

1. Introduction 57
2. How should an ideal carrier behave for gene delivery? 59
3. Basic molecular structure of polymeric carriers for non-viral gene delivery 62
4. Structural improvement of polymeric vectors for gene delivery 65
   4.1. Shielding of the polyplexes 65
   4.2. Shaping polyplexes on preformed templates 68
   4.3. Functionalizing polyplexes with ligands 69
5. What could be done for improving the situation? 70
6. Summary and last remarks 72
Acknowledgements 72
References 72

1. INTRODUCTION

Gene therapy possesses an enormous potential for successful treatment of several genetic-based or acquired major diseases [1]. Thus, the field of gene therapy has been extensively investigated during the last decades. However, the main drawback for successful gene therapy remains the efficient *in vivo* delivery of the genetic material into the nucleus of specific cells targeted for therapy [1]. Naked, uncomplexed DNA is not able by itself to neither survive the harsh conditions of the human bloodstream nor undergo a transfection process under physiological conditions (i.e. crossing cell membranes and penetrating nucleus for being expressed by transcription). Thus, the development of efficient DNA carriers (gene delivery) is one of the key points of gene therapy [2].

Classically, the carriers for gene delivery can be divided into two main families: viral carriers and non-viral carriers [3]. Up to now, the first family was undoubtedly the most explored and was used in most of the clinical trials [4]. In this case, modified viruses (typically retroviruses or adenoviruses) are used to transport intracellularly the genetic material. Such systems were found to be highly effective for *in vitro* and *in vivo* gene delivery (transfection efficiency is usually close to 100%) but their definitive clinical adoption is still threatened by major toxicity risks. In particular, two incidents happening during clinical trials in 1999 and 2000, restrained the use of viral vectors [5,6]. As a consequence, it also considerably boosted the up-to-then relatively marginal field of artificial gene delivery vectors (the term "non-viral" appeared in the literature in the mid-1990s and clearly reflects that this field has emerged in opposition to viral systems). Indeed artificial (non-viral) vectors based on lipids, natural polymers or synthetic polymers show much lower safety risks and therefore received full attention from the scientific community after the year 2000 [2]. However, toxicity is definitely not the only factor, which justified the research on non-viral gene carriers (although the toxicity issue was often used as the main argument in recent literature on non-viral vectors). Indeed viral vectors also suffer from other important drawbacks such as very high cost, low availability and low DNA storage capability. These limitations could be potentially overcome by using synthetic gene carriers.

In particular, synthetic polymeric gene carriers were particularly studied recently since these systems possess all the inherent advantages of synthetic polymers: they are usually relatively cheap (although this kind of specialty materials are indeed more expensive than standard commodity polymers) and can be generally prepared in high yields *via* facile synthetic methods. In this regard, synthetic polymers capable of transporting *in vitro* or *in vivo* genetic material constitute a very important class of advanced materials [7].

The first examples of polymeric gene carriers were reported in the early 1990s [8–15]. Afterwards, the field grew exponentially. Several types of interesting polymeric DNA carriers were designed, prepared and characterized during the last 10 years. However, up to now, the performances of these artificial systems are still far from ideality. The purpose of the present chapter is to review recent progress in the field of polymeric gene carriers and to discuss the key aspects, which could lead to the construction of high-performance polymeric gene carriers.

2. HOW SHOULD AN IDEAL CARRIER BEHAVE FOR GENE DELIVERY?

The delivery of genetic material into the nucleus of a targeted cell is a difficult task hindered by multiple biological barriers. The mechanisms of transfection are still not fully understood and indeed vary, depending on the type of studies (*in vitro* or *in vivo)*, on the route of administration of the gene carrier, on the nature of the carriers and on the variety of targeted cells. Thus, a detailed description of the mechanisms of transfection is beyond the scope of this chapter and can be found in the previous chapter of this book or several recent reviews [7,16,17]. In brief, a simplified description of the transfection process is presented in Scheme 1. For transporting DNA in a cell nucleus, a gene carrier should first travel in a physiological media (the bloodstream *in vivo* or a physiological solution *in vitro*), cross the membrane of a cell, travel intracellularly in the cytoplasm and finally release the gene material inside or in close proximity of the cell nucleus. Hence, for completing such a demanding mission, an ideal carrier for gene delivery should fulfill numerous drastic requirements.

Table 1 lists the main prerequisites for successful gene delivery. This list is not exhaustive since as mentioned above, the required properties of a gene carrier might differ from study to study. However, Table 1 gives a general impression of necessary conditions for an optimal transfection. Indeed, the list is quite long, which clearly illustrates that finding an ideal carrier for

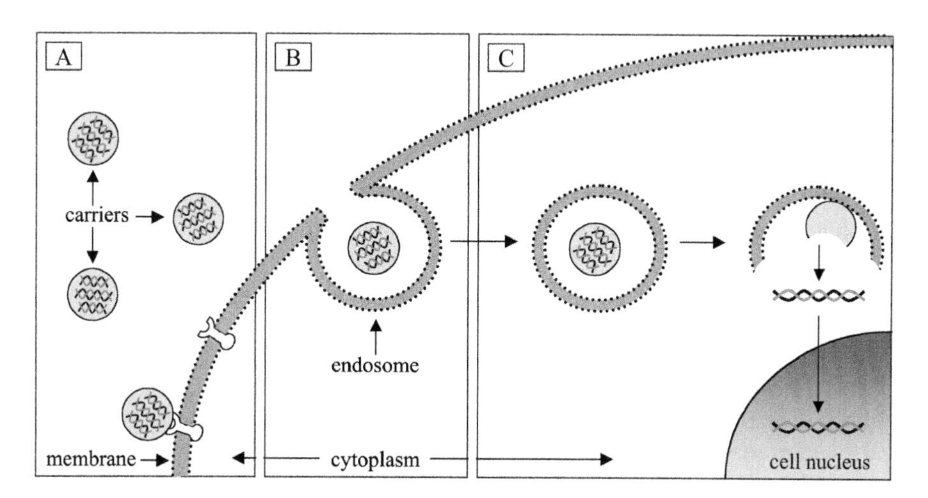

Scheme 1. Simplified representation of a transfection process: (A) adhesion of the DNA carriers to cell surface receptors, (B) internalization of the carrier *via* receptor-mediated endocytosis, (C) release of the genetic material in the cytoplasm and nuclear trafficking.

Table 1. List of fundamental requirements for an ideal gene carrier

	Structural requirements
1	Ability to contain DNA
2	Colloidal stability in physiological media
3	Size in the nanometer range
4	High transport capability
	Biological requirements
5	Resistance to immune system
6	Resistance to plasma protein adsorption
7	Non-toxic
8	Biodegradable
9	Ability to cross cellular membranes
10	Ability to escape endosomes and to release DNA
11	Nuclear localization and entry
12	Cell specificity

gene delivery is a challenging objective. First of all, an optimal gene carrier should fulfill precise structural requirements. The primary fundamental condition is that the structure should be able to contain and transport large quantities of DNA (Table 1, entries 1 and 4). A good carrier should be able to bind DNA macromolecules either on its surface or in a specific inner region (core, segregated domain, layer). Moreover, this entrapment should be based on weak interactions (e.g. electrostatic interactions, van der Waals forces, labile covalent bonds) in order to allow later an intracellular release of the genetic material (Table 1, entry 10). As a second fundamental requirement, the carrier should be colloidally stable in physiological conditions (Table 1, entry 2) in order to not precipitate or agglomerate extracellularly. Nevertheless, the size of the carriers is also a factor of importance (Table 1, entry 3). For instance *in vivo*, an efficient gene carrier should typically possess an average size below a few hundred nanometers in order to not be instantly detected and engulfed by macrophages (Table 1, entry 5). On the other hand, the particulates used as carriers should be big enough to contain significant quantities of DNA (Table 1, entry 4). Thus, ideal size should be balanced between lower and upper limits (typically, the ideal size is assumed to be in the range 100–200 nm).

Besides these structural conditions, an ideal carrier for gene delivery should also possess several biological properties. First, it should resist the various defense mechanisms of the host organisms (Table 1, entry 5) either extracellularly (e.g. macrophages *in vivo*) or intracellularly (e.g. lysosomes). In particular, the carrier should be able to resist *in vivo* to surface adsorption of specific plasma proteins such as immunoglobulins (Table 1, entry 6), which can trigger immune response. Moreover, it should be non-toxic and rapidly degraded after delivery process (Table 1, entries 7 and 8). Another fundamental biological requirement for successful transfection is of course that the gene carrier should efficiently cross cell bilayer lipid membranes (Table 1, entry 9). Nevertheless, if the carrier enters cells *via* endocytosis (Scheme 1), it should be able to escape from the endosome and release the load (naked DNA) inside the cell (Table 1, entry 10). The release should indeed be done inside (or at least in close proximity) of the cell nucleus in order to allow intranuclear expression of the DNA (Table 1, entry 11). Last but not the least, for successful gene therapy *in vivo*, an ideal gene carrier should be eventually able to target specific cells or organs inside the body (Table 1, entry 12).

It appears at first sight from Table 1 that finding an ideal carrier for gene delivery is a particularly hard to reach target. One good option is to use modified biological molecules as gene-carriers, which are known to possess the required structural and biological properties. In that regard, modified viruses (viral vectors) were obvious candidates for gene delivery since they perform in nature spontaneous transfection processes [1,4,18]. Indeed, they invade host cells to deliver their own genetic material in the cell nucleus and thus, by essence, fulfill the requirements 1, 2, 3, 5, 6, 9, 10 and 11 of Table 1. Hence, it was verified experimentally that modified viral vectors exhibit a very high transfection efficiency for DNA delivery [4]. However, as mentioned in the introduction, they also present significant limitations concerning DNA storage capability and safety [18]. Therefore, although they are very promising candidates, viral vectors cannot yet be considered as "ideal" gene carriers. Therefore, another option is to construct artificial carriers, which exhibit all the properties listed in Table 1. The latter is indeed much easier to conceptualize than to implement. However, this ultimate goal might be reached if smart synthetic strategies are used. One first approach is the "bottom-up" spontaneous self-assembly of molecular building-blocks [19]. In this case, the molecular building-blocks are thought and designed for solving both structural and biological issues. The building-blocks can be prepared using various synthetic chemistry techniques and can potentially be either synthetic macromolecules, low molecular-weight synthetic molecules, modified biomolecules or hybrids of synthetic and biological molecules. Afterward, the building-blocks are self-assembled in physiological media in a predicted nanostructure, which can be used as nanocontainers for delivery (such self-assembly processes are typically driven by various types of supramolecular interactions such as van der Waals forces, π–π interactions, electrostatic interactions or H-bonding) [20]. As a complement to a bottom-up self-assembly approach, a "biomimetic" strategy may also be a key to successful design of high-performance gene carriers [21]. Indeed, one can learn from Nature and use specific molecular moieties of natural systems for functionalizing synthetic building-blocks. For example, specific molecular fragments of viruses, which are responsible for their unusual cell penetration capability, can be used to functionalize artificial systems [22]. Such virus-like artificial systems (virus-like but non-viral) could be in the future the key to successful gene delivery. In the next two paragraphs, the molecular solutions which were studied up to now for preparing efficient polymeric building-blocks for gene carriers are reviewed.

3. BASIC MOLECULAR STRUCTURE OF POLYMERIC CARRIERS FOR NON-VIRAL GENE DELIVERY

Polymer science is a powerful synthetic platform, which offers numerous cheap practical solutions to multiple applicative problems. For instance, synthetic polymer systems have been widely studied and applied in biotechnology and medicine [17,23–25]. The latter is in part due to the extreme versatility of modern synthetic polymer chemistry, which allows preparation of tailor-made macromolecules (i.e. polymers with controlled chain-length, architectures, composition and functionalities) specifically designed for targeted applications [26–28]. Therefore, polymer chemistry was indeed also a very tempting tool for preparing specific carriers for the very demanding field of non-viral gene delivery. However, as discussed in the previous paragraph, an ideal polymeric carrier for gene delivery should fulfill various drastic requirements. Up to now, carriers based on synthetic polycations are the most promising polymeric materials for overcoming all the aforementioned barriers [7,29]. Polycations are macromolecules, which possess multiple cationic sites in their molecular structure at physiological pH. Thus, these positively charged polymers undergo electrostatic interactions with the negatively charged backbone of DNA (phosphate groups). The latter results in the formation of compact aggregates polycations/DNA (commonly named polyplexes) possessing average sizes in the range 10–200 nm. The shape (or morphology) of the formed polyplexes depends on multiple parameters (chemical structure of the polycations, molecular weight of the polycations, DNA/polycation ratio, preparation conditions for assembling DNA and polycation...) and thus can be variable (i.e. toroid, globular or rod-like). However, this compaction driven by electrostatic interactions leads to a spectacular decrease of the volume of giant molecules of DNA and therefore significant amounts of genetic material can be contained in a single polyplex [30]. Hence, aggregates of polycations and DNA fulfill structural requirements 1, 3 and 4 of Table 1 and therefore could be used as vectors for non-viral gene delivery. Moreover, polyplexes are usually positively charged (an excess of positive charges as compared to negative charges of DNA is often required for maximum compaction) and therefore additionally fulfill some biological requirements of Table 1. For instance, positively charged polyplexes spontaneously interact with negatively charged proteoglycans located on cell membranes and subsequently penetrate cells *via* endocytosis (Table 1, entry 9) [31,32].

Moreover, intracellularly, it was shown that polyplexes are able to escape the endosomes (Table 1, entry 10). A main theory known as the "proton-sponge" was proposed for explaining this endosomal release [33,34]. This hypothesis suggests that polyplexes, which are only partially protonated in physiological conditions become more protonated in the acidic interiors of the endosomes (Scheme 2). The latter first results in an influx of counterions in the endosomes and subsequently in an osmotic swelling of the endosome, which ultimately leads to a rupture of the endosome and a liberation of the polyplex in the cytoplasm. The proton-sponge theory was confirmed by experimental evidences [33,34], although this effect is more or less significant depending on the type of polycation used for building the polyplex [35,36]. Nevertheless, it was evidenced that polyplexes promote the entry into the cell nucleus (Table 1, entry 11) [32,37]. The mechanisms involved are still unclear but seem to be more a consequence of the three-dimensional structure of the polyplexes rather than their surface charge [37].

In the last 10 years, various types of polycations were reported to be interesting for applications in non-viral gene delivery. All these promising macromolecules are polyamines (i.e. polymers bearing multiple amine groups). Under physiological conditions, the amine moieties of these polymers are protonated and therefore electrostatic interactions with DNA occur. The architecture (molecular three-dimensional structure) of synthetic polyamines can be of different types such as linear, grafted, hyperbranched or dendritic. Scheme 3 shows the molecular structure of the most studied polycations in gene delivery. Historically, the first studied polyamine for gene delivery was poly(L-lysine). This polypeptide can be prepared either in a very defined way by Merrifield solid-phase peptide synthesis or in a more straightforward way *via* polymerization of *N*-carboxy anhydrides [38–40]. Poly(L-lysine) has

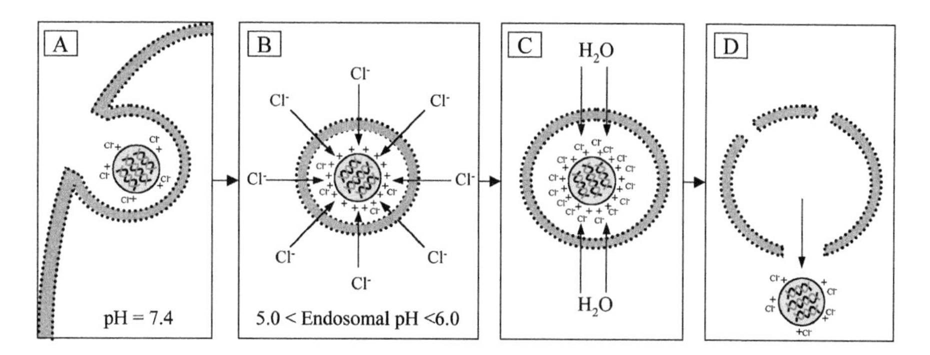

Scheme 2. Schematic description of the "proton sponge" theory: (A) polyplex in physiological conditions, (B) buffering effect of the polyplex in the acidic interior of an endosome, (C) osmotic swelling of the endosome, (D) endosomal rupture and escape of the polyplex.

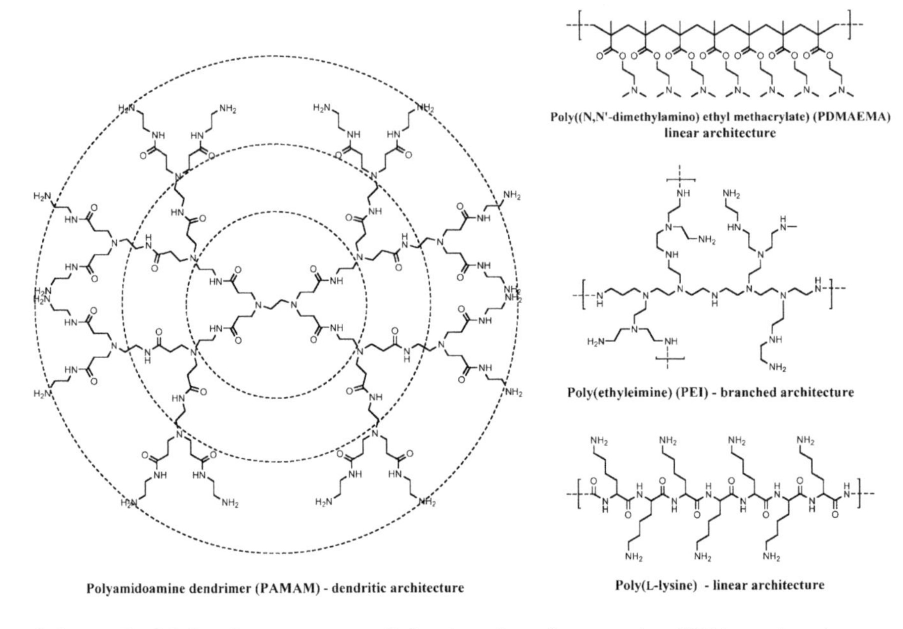

Scheme 3. Molecular structures of classic polycations used as DNA carriers in non-viral gene delivery.

been widely studied in gene delivery either as a linear polymer or as a dendrimer [41,42]. Another classic polyamine extensively studied in non-viral gene delivery is polyethyleneimine (PEI) [43]. This polymer was first investigated by Behr and coworkers [15] and later proven to be the most efficient polymeric carrier for gene delivery [43]. Thus, a considerable amount of studies have been reported on PEI/DNA polyplexes. This large field was recently exhaustively reviewed by Göpferich and coworkers [43]. Besides poly (L-lysine) and polyethyleneimine, several other synthetic polyamines have been studied for application in non-viral gene delivery such as poly((*N,N'*-dimethylamino) ethyl methacrylate) (PDMAEMA) [44,45], poly(3-guanidinopropyl methacrylate) [46], poly(amidoamine) dendrimers (PAMAM) [13], poly(amidoamine) linear derivatives [47], aminated poly(L-histidine) [48], aminated poly(L-glutamic acid) [49], polyspermine [50], poly-L-ornithine [51] or poly(*N*-ethyl-4-vinylpiridinium) bromide [10]. Most of these polymers have been only marginally studied but some, such as PDMAEMA and PAMAM have been more regularly investigated and proven to be interesting candidates for gene delivery.

Non-viral gene vectors based on synthetic polyamines allow intracellular delivery of genetic material (transfection process). Several *in vitro* studies confirmed this behavior for all the polyamines listed above. However, although a true transfection process can be observed, polymeric non-viral gene delivery systems all exhibit a significant lack of efficiency (*in vitro* and moreover *in vivo*) as compared for example to viral vectors (it is usually assumed that polymeric gene carriers are a 1000 times less efficient than viral vectors). The latter is first a consequence of the often very poor aqueous stability of polyplexes of polyamine/DNA, but moreover of the long and dangerous journey that the nanocarriers experience either *in vitro* or *in vivo* for reaching cell nucleus (Scheme 1). Along their way to the nucleus, polyplexes at first and ultimately naked DNA both encounter multiple extracellular or intracellular obstacles, which all lead to significant loss of genetic material. Extracellularly (*in vivo* case only), as mentioned in the first paragraph, the main problem is to resist to the harsh environment of body fluids, where polyplexes can be easily cleared out *via* immune mechanisms. Nevertheless, intracellularly, multiple additional problems arise. First major limitation is indeed the cellular uptake of the polyplexes. Although the real mechanisms of endocytosis (including uptake and endosomal escape) are still not fully understood, it is unanimously believed that this step leads to a significant destruction of the transported genetic material. In particular, the low pH environment in the endosomes as well as the enzymatic degradation due to endosome/lysosome fusion are two possible reasons of low efficiency in polycationic gene delivery. Moreover, assuming that a polyplex can survive endocytosis and successfully escape endosomes, the released material should also survive to the nucleases present in the cytosol before reaching its ultimate target in the cell nucleus. In addition to very low transfection efficiency, polycationic non-viral gene delivery vectors also often exhibit cytotoxicity (even though they are typically much safer than viral systems). Hence, on the whole, polymeric gene carriers based on simple polyamine/DNA aggregates are far from being ideal materials for gene delivery.

4. STRUCTURAL IMPROVEMENT OF POLYMERIC VECTORS FOR GENE DELIVERY

4.1. Shielding of the polyplexes

As mentioned above, polyplexes formed only by a polycation and DNA often suffer several limitations such as low water solubility, poor colloidal stability (aggregation behavior) and strong interactions with serum proteins (i.e. poor resistance to *in vivo* environment). In order to overcome such limitations, the structure of the polyplexes should be modified. For instance, the incorporation of poly(ethylene oxide) (PEO) (also known as poly(ethylene glycol) (PEG) depending on the molecular weight) in the structure of the carriers was proven to be an efficient method for drastically improving the properties of polymeric non-viral gene delivery systems (Scheme 4). PEO is

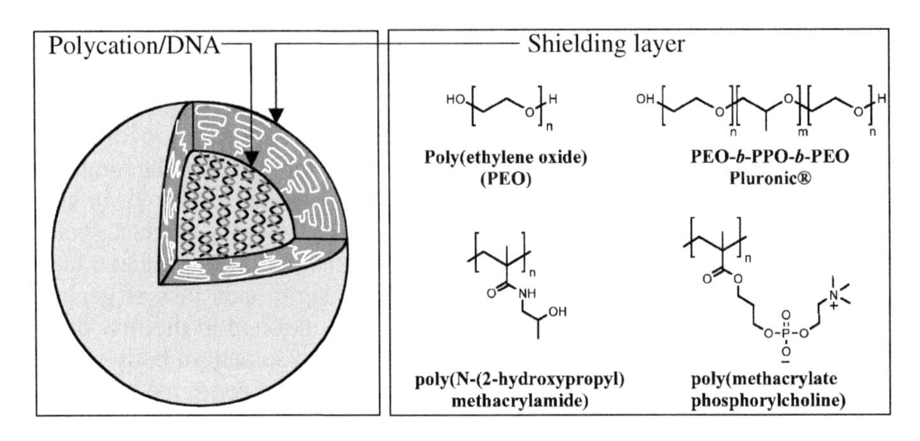

Scheme 4. Simplified representation of a protected polyplex and molecular structure of polymers used for shielding.

a cheap, water soluble, biocompatible, FDA approved polymer, which is therefore used in various aspects of biotechnology and medicine [52–54]. Modification of polyplexes with PEO (or PEGylation) can be done *via* two main synthetic pathways. First approach relies on the molecular modification of the polycation building-blocks (pre-PEGylation) before their association with DNA (a typical example of bottom-up self-assembly approach). In the second case, preformed polyplexes are directly functionalized by poly(ethylene oxide) (post-PEGylation). Although the latter pathway was successfully used in very significant work [55], this approach is very limited. First, post-PEGylation only allows a rough control over polyplexes morphology. Nevertheless, the resulting polyplexes are difficult to characterize and therefore problematic to use for additional molecular modifications. On the other hand, pre-PEGylation is a much more versatile synthetic pathway, which allows a precise control on the molecular structure of the building-blocks (i.e. the molecular structure of the blocks can be carefully analyzed *via* standard analytical methods before complexation with DNA) and therefore on the polyplexes structure. Thus, in most of the reported studies pre-PEGylation approaches were used [43,56].

Most of the polycations mentioned in the previous paragraph have been coupled to PEO segments and the resulting PEO-polycation conjugates have been tested in gene delivery. For instance various types of chemistry have been used for PEGylation of PEI [43,57–61]. Overall, the most used method is the direct coupling reaction of preformed functional PEO and PEI segments, which is indeed efficient but also a quite limited route for controlling the molecular structure of the resulting conjugates [43]. A very different approach is typically used for constructing conjugates of PEO and poly (L-lysine). In most cases, block copolymers PEO-*b*-PLL are synthesized *via* the ring-opening polymerization of the *N*-carboxy anhydride derivative of L-lysine, initiated by an amino functional PEO macroinitiator [56,62–65]. This method allows typically a tailor-made synthesis of well-defined PEGylated polycations [39,40,56,66]. Moreover, this technique allows a very careful control over chain-end functionality, which opens the possibility to functionalize further the conjugates. The recent works of Kataoka and coworkers elegantly illustrates the versatility of this synthetic approach [62,65]. Similar synthetic strategies employing macroinitiators were also used for constructing PEGylated derivatives of PAMAM and PDMAEMA. In the case of PAMAM, dendrimers were grown from amino functional PEO precursors *via* a stepwise synthesis [67]. PEO-*b*-PDMAEMA conjugates were synthesized using PEO macroinitiators either *via* group transfer polymerization or anionic polymerization [68,69].

The incorporation of PEO in polyplexes changes in many ways their structural and biological properties. First, the aqueous colloidal stability of PEGylated polyplexes is significantly improved as compared to non-PEGylated polyplexes [55,68]. PEGylated polycations form with DNA small stable nanoparticles in aqueous medium (even though the presence of PEO in pre-PEGylated systems hinders to some degree the complexation of the polycation and DNA). Aggregation behavior of these nanoparticles is consequently reduced as compared to standard polyplexes polycation/DNA [55,68]. Additionally, the resistance of polyplexes to defense and degradation mechanisms is drastically increased when PEO is used as a shielding layer. PEGylation was proven to efficiently prevent polyplexes from plasma protein adsorption [55], to increase their *in vivo* circulation time in the bloodstream [55,70] and to protect them from nuclease attacks [71]. Nevertheless, a lower cytotoxicity was often observed *in vitro* for PEGylated polyplexes as compared to uncoated ones [61,63]. The influence of PEGylation on transfection efficiency depends on several experimental parameters (chain length of the PEO segment, composition PEO/polycation) and therefore results vary from study to study. The presence of PEO modifies the overall surface charge of the gene carriers, thus it is expectable that PEGylation might reduce to some degree the cellular uptake of polyplexes [56]. Recent studies effectively highlighted limited *in vivo* performances for PEGylated polyplexes [70,72,73]. However, altogether PEGylated polyplexes fulfill most of the requirements of Table 1 (requirements 1–11 are completed). The latter illustrates that simple structural modifications of the polyplexes (or modifications of their building-blocks precursors) can tune dramatically their properties. Thus, PEGylation is from now on more of a requirement than an option for optimal non-viral gene delivery.

Besides poly(ethylene oxide), other possibilities have been studied for polyplexes shielding. For instance other polymeric materials such as poly(acrylic acid) [74], poly(methacrylate phosphorylcholine) [75], poly(*N*-(2-hydroxypropyl)methacrylamide) [76], *N*-isopropylacrylamide [77] or triblock copolymers PEO-*b*-poly(propylene oxide)-PEO [78,79] (known commercially as Pluronic®) have been used to stabilize polyplexes (Scheme 4). Although some of these studies were quite promising, most of these shielding alternatives have been only marginally studied. Plasma proteins such as transferrin [80] or human serum albumin [81] have also been studied as protecting external layer for polyplexes. These approaches were quite successful since plasma proteins are by essence biocompatible and constitute a good protection against immune defenses.

4.2. Shaping polyplexes on preformed templates

Typically polyplexes are nanoparticles formed *via in situ* electrostatic self-assembly of DNA and polycations. As mentioned above, the morphology and colloidal stability of these particles vary depending on several experimental parameters. One strategy for controlling more precisely the morphology (size, shape, polydispersity) of polymeric gene carriers is to build the polyplexes on preformed defined nanoparticles. For instance, polymeric micelles, polymer micro(nano)spheres or inorganic particles can be used as a template for the polyplexes. Polymeric micelles are usually obtained *via* the aqueous self-assembly of macrosurfactants (i.e. copolymers possessing distinct hydrophilic and hydrophobic segments) [82]. In most cases micelles are spherical nanoparticles composed of a hydrophobic core and a hydrophilic shell [20,82]. These objects can be used as gene delivery carriers if the outer-shell is made of a polycation. For example spherical micelles obtained *via* the aqueous self-assembly of polyethyleneimine-*graft*-poly(methyl methacrylate) [83] (PEI-*g*-PMMA) or poly(L-lysine)-*graft*-poly(lactide-*co*-glycolide) [84] (PLL-*g*-PLGA) were tested as non-viral vectors. The hydrophilic polyamine shells of the micelles (i.e. PEI or PLL) are able to condense and transport DNA, whereas the segregated hydrophobic cores provide cohesion to the structure (PMMA or PLGA are glassy polymers, thus micellar cores are "frozen") [83,84]. Similar types of nanocarriers were also prepared using nanospheres. Polymeric micro(nano)spheres are usually prepared *via* dispersion techniques in the presence of surfactants. For instance, biodegradable microspheres of poly(lactide-co-glycolide) (PLGA) were extensively applied as drug-delivery systems. Thus, such particles were also tested in non-viral gene delivery [85,86]. PLGA nanospheres possessing a PEI layer on their surface were reported to efficiently complex DNA and to allow transfection [85,86].

Polymer-based gene vectors were also constructed on inorganic cores. For example, Przybycien and coworkers investigated colloidal gold–PEI conjugates as vectors for gene delivery [87]. Plank and coworkers coated the surfaces of superparamagnetic iron oxide nanoparticles with PEI [88]. The resulting hybrid organic–inorganic nanoparticles were used for transporting DNA either *in vitro* or *in vivo*. Moreover, the presence of a superparamagnetic core in the particle allows to force transfection *via* the use of external magnetic fields (magnetofection) [88]. This new method was proven to be a very efficient and promising tool for targeted gene delivery. Very recently, Liu and coworkers tested PEI grafted on the surface of multiwalled carbon nanotubes as a new vector for non-viral gene delivery [89]. They measured a high transfection for these objects and proposed that the nanotubes template could favor the endosomal escape of the polyplexes *via* an enhanced proton-sponge mechanism.

4.3. Functionalizing polyplexes with ligands

In comparison to viral vectors, artificial polyplexes possess generally a very low transfection efficiency. Moreover, shielding the polyplexes with PEO or other protecting layers usually leads to reduced performances (even though shielding layers are very beneficial in terms of physiological stability *in vitro* and *in vivo*). Hence, additional structural improvements of the polymeric carriers might be capital for optimal performances. In particular, conjugation with biomolecules permitting organ targeting, cell targeting, cell penetration, endosomal escape or nucleus targeting is of particular importance. Interesting progress have been already made in this direction [7,22,41,56,90].

Several pathways have been investigated to promote cellular internalization. A first strategy relies on receptor-mediated endocytosis. Polyplexes can be modified with receptor ligands for increasing binding to cell surface receptors. This approach not only promotes cellular entry but is also cell specific, and thus opens wide possibilities in cell targeting. For example, folic acid (recognized by the folate receptor) [91,92], transferrin (recognized by the transferrin receptor) [55,93], the tripeptide RGD (recognized by the integrin receptor) [94], galactose (recognized by the asialoglycoprotein receptor) [95,96] or mannose (recognized by the mannose receptor) [97,98] have been used for enhancing cell entry of polyplexes. Another approach for promoting cellular internalization is to conjugate synthetic polycations and peptide sequences found in viruses (cell-unspecific approach). For instance, specific sequences known as protein tranduction domains (PTD) were proven to promote very efficiently intracellular entry [99–101]. In particular, TAT peptide (a specific sequence of HIV-1 virus) was tested in non-viral gene delivery and proven to be very efficient [7,102]. However, although promising, the conjugation with PTD is still rather rare in polycation-based gene delivery.

Specific peptide sequences from viruses can also be used for favoring endosomal escape. Indeed, viruses possess fusion peptides (or membrane disruption peptides) on their surface, which allow them to leave endosomes *via* membrane destabilization. Peptides such as INF7 (a domain of the influenza virus), GALA (synthetic sequence) or KALA (synthetic sequence) exhibit fusogenic activities with phospholipid membranes in acidic endosomal conditions [7,103]. The biomimetic functionalization of polyplexes with such sequences was effectively proven *in vitro* to be a very effective pathway to promote endosomal escape and therefore to increase transfection efficiency [46,103].

The transfection process can also be dramatically improved by using nuclear localization peptide sequences (NLS). A plethora of NLS has been characterized in the past and several of them have been used for modifying polyplexes [90]. More information about the use of NLS in non-viral gene delivery can be found in the chapter 4 of this book.

On the whole, the modification of polyplexes with biologically active moieties was proven to be in many cases very promising. However, such synthetic modifications often require to work at the interface between polymer science and biochemistry (e.g. at the interface between synthetic polymer chemistry and peptide chemistry) [104]. The latter is somewhat a very young field of research, thus more efforts are needed in this direction.

5. WHAT COULD BE DONE FOR IMPROVING THE SITUATION?

The previous paragraphs highlighted that, although polymeric non-viral gene carriers are a very interesting alternative to viral vectors, up to now their performances and thus their medical potential are quite limited. However, as discussed above, their properties can be noticeably improved by modifying their molecular structure. Each *in vitro* or *in vivo* limitation can be potentially overcome by using tailor-made macromolecular designs. For instance, the incorporation of poly(ethylene oxide) in the carriers structure helped in preventing several obstacles (e.g. better aqueous stability of the polyplexes, protection from the immune system, lower toxicity, protection from nucleases...). More generally, it is reasonable to assume that every *in vitro* or *in vivo* problem probably has its molecular solution. Therefore, it seems clear that the upper performance limit of polymeric gene vectors was not reached yet. However, for preparing ideal polymeric carriers that could circumvent all known limitations, a deep understanding of the relationship between structure and properties for polymeric vectors is crucial. In this regard, a plethora of major questions arise: what is the influence of the structure of the modified polycations (architecture, molecular weight, polydispersity, composition, functionality) on the shape of the polyplexes and on their behavior? What is the influence of the mixing ratio polycations/DNA on the shape of the polyplexes, on their stability in aqueous medium and on their intracellular dissociation? What is the optimal content of poly(ethylene oxide) needed for the best stability *in vivo*? What is the influence of the morphology of the polyplexes on the cellular uptake? What is the best way to reduce carrier toxicity? What would be the best molecular modifications for promoting cell targeting, membrane penetration, endosomal escape or nucleus targeting? Several studies have already been made for solving most of these questions [105–109]. Hence, some of them have been already partially answered. However, it seems more and more that the accurate answers to all these questions require a multidisciplinary approach. For instance, the preparation of tailor-made polymeric non-viral gene carriers necessitates more and more skills on macromolecular synthesis and is therefore mostly the responsibility of specialists such as polymer chemists, biochemists and material scientists. Additionally, the interactions between synthetic polycations and DNA as well as the study of the resulting polyplexes morphologies are typically topics at the interface of polymer physics and colloidal science. Moreover, *in vitro* and *in vivo* tests of non-viral gene delivery systems require high skills and knowledge in physiology, pharmaceutical sciences and medicine. Some research groups in the past have already created solid bridges between these various aspects of material science and medicine. However, in the future, it seems more and more evident that the practical keys to successful gene delivery will not come from isolated research groups but from strongly interconnected multidisciplinary research networks. In that context, more inter-departmental collaborations in the field of non-viral gene delivery are certainly needed. Moreover, for fruitful collaborations, the various actors and research groups involved should to some degree "speak the same language". The latter is not obvious, since the short-term goals of the very different scientific communities of material science and medicine are often very distant. For example, the common holy grail in contemporary material science is the search for nanostructuration. The objectives in this "nano world" are often clearly more related to design than properties. The modern material scientist is more of an architect than an engineer. Thus, from a material scientist point of view, an organized regular nanostructure is often more valuable scientifically than an awkward efficient aggregate. The search for properties in material science is often more a justification than a real goal. On the other hand, performances and properties are clearly the main targets of pharmaceutical and medical teams involved in gene therapy research. Therefore, for an efficient screening of the multiple structure/properties parameters influencing non-viral gene delivery, communication, understanding and harmony between scientific lobbies are undoubtedly needed. In order to reach this goal, strong existing barriers should be torn down. A good example of the partition of scientific community is the separation of literature data. Sharing knowledge is important for creating interdisciplinary bridges. However, a relevant article in material science will be most likely published in journals such as "*Journal of the American Chemical Society*" [110] or "*Advanced Materials*" [111], which are probably poorly read by the medical community. Similarly, relevant medical publications are probably out of sight for the material science community. A journal such as "*Bioconjugate Chemistry*" [112] is probably one of the rare examples of a periodical publication where communities meet. Therefore, for progress in non-viral gene delivery, more interdisciplinary knowledge interfaces have to be built.

Besides multidisciplinarity, another problem is indeed the competitiveness in the field of gene delivery. Although competition is definitely needed for stimulating progression, it also results in the overproduction of incomparable data. Every research group wants to claim a better system than the others (classic criteria are a high transfection and a low toxicity). For example, in the restricted case of polycationic non-viral gene delivery, based on available literature data, it is almost impossible to simply say what synthetic polymer overrules the others. Most likely, all systems present interesting advantages, depending on the specific kind of targeted application. But the fact is that the competitive situation generates a lack of careful comparison of all reported systems. Such a situation should indeed be improved *via* a unification of knowledge. In this regard, new combinatorial approaches might be interesting tools for rationalizing the gigantic field of non-viral gene delivery [113].

6. SUMMARY AND LAST REMARKS

The field of polymeric non-viral gene delivery grew exponentially during the last decade. Several interesting polymer-based gene vectors have been proposed in the literature and tested medically. However, on the whole, all the reported systems still exhibit unsatisfying performances. For improving this situation, both the molecular structure and the nanostructure of the polymeric vectors should be adapted to the biological problems encountered either *in vitro* or *in vivo*. The latter is probably only possible if a multidisciplinary research force is created. In the future, specific actions (e.g. congresses, workshops, publications, creation of data libraries) should be taken for promoting and unifying the field of non-viral gene delivery.

ACKNOWLEDGEMENTS

The Deutsche Forschungsgemeinschaft (DFG) is greatly acknowledged for supporting our research in the field of non-viral gene delivery (LU 1195/1-1). Moreover J.F.L. thanks Ali Ben Salem for organizing the Berlin international workshop on gene therapy and Dr. Régis Cartier for stimulating discussion.

REFERENCES

[1] I. M. Verma and N. Somia, *Nature*, 1997, **389**(6648), 239–242.
[2] D. Luo and W. M. Saltzman, *Nat. Biotechnol.*, 2000, **18**(1), 33–37.

[3] K. Lundstrom and T. Boulikas, *Technol. Cancer Res. Treat.*, 2003, **2**(5), 471–485.
[4] K. Lundstrom, *Trends Biotechnol.*, 2003, **21**(3), 117–122.
[5] E. Marshall, *Science*, 1999, **286**(5448), 2244–2245.
[6] S. Hacein-Bey-Abina, C. Von Kalle, M. Schmidt, M.P. McCcormack and N. Wulffraat, *et al.*, *Science*, 2003, **302**, 415.
[7] T. Merdan, J. Kopecek and T. Kissel, *Adv. Drug Del. Rev.*, 2002, **54**(5), 715–758.
[8] G. Y. Wu and C. H. Wu, *J. Biol. Chem.*, 1987, **262**, 4429–4432.
[9] E. Wagner, M. Zenke, M. Cotten, H. Beug and M. L. Birnstiel, *Proc. Natl. Acad. Sci. USA*, 1990, **87**, 3410–3414.
[10] A. V. Kabanov, I. V. Astaceva, M. L. Chikindas, G. F. Rosenblat, V. I. Kiselev, E. S. Severin and V. A. Kabanov, *Biopolymers*, 1991, **31**, 1437–1443.
[11] E. Wagner, M. Cotten, R. Foisner and M. L. Birnstiel, *Proc. Natl. Acad. Sci. USA*, 1991, **88**, 4255–4259.
[12] M. Cotten and E. Wagner, *Curr. Opin. Biotechnol.*, 1993, **4**, 705–710.
[13] J. Haensler and F. C. Szoka, *Bioconj. Chem.*, 1993, **4**(5), 372–379.
[14] F. D. Ledley, *Curr. Opin. Biotechnol.*, 1994, **5**, 626–636.
[15] O. Boussif, F. Lezoualc'h, M. A. Zanta, M. D. Mergny, D. Scherman, B. Demeneix and J. Behr, *Proc. Natl. Acad. Sci. USA*, 1995, **92**, 7297–7301.
[16] C. W. Pouton and L. W. Seymour, *Adv. Drug Del. Rev.*, 2001, **46**, 187–203.
[17] B. Twaites, C. de las Heras and C. Alexander, *J. Mater. Chem.*, 2005, **15**, 441–455.
[18] C. E. Thomas, A. Ehrhardt and M. A. Kay, *Nat. Rev. Genet.*, 2003, **4**, 346–358.
[19] R. S. Tu and M. Tirrell, *Adv. Drug Del. Rev.*, 2004, **56**(11), 1537–1563.
[20] S. Förster and T. Plantenberg, *Ang. Chem. Int. Ed.*, 2002, **41**(5), 688–714.
[21] C. Sanchez, H. Arribart, M. M. Giraud Guille, *Nat. Mater.*, 2005, **4**, 277–288.
[22] L. D. Shea and T. L. Houchin, *Trends Biotechnol.*, 2004, **22**(9), 429–431.
[23] R. Freitag (ed.), *Synthetic Polymers for Biotechnology and Medicine*, Landes Bioscience, Georgetown, 2002, p. 164.
[24] R. Duncan, *Nat. Rev. Drug Discovery*, 2003, **2**, 347–360.
[25] R. Langer and D. A. Tirrell, *Nature*, 2004, **428**, 487–492.
[26] N. Hadjichristidis, M. Pitsikalis, S. Pispas and H. Iatrou, *Chem. Rev.*, 2001, **101**, 3747–3792.
[27] J. Pyun, X.-Z. Zhou, E. Drockenmuller and C. J. Hawker, *J. Mater. Chem.*, 2003, **11**, 2653–2660.
[28] D. Shipp, *J. Macromol. Sci., C Polym. Rev.*, 2005, **45**, 171–194.
[29] S. C. De Smedt, J. Demeester and W. E. Hennink, *Pharmaceutical Res.*, 2000, **17**(2), 113–126.
[30] V. A. Bloomfield, *Curr. Opin. Struct. Biol.*, 1996, **6**, 1471–1481.
[31] K. A. Mislick and J. D. Baldeschwieler, *Proc. Natl. Acad. Sci. USA*, 1996, **93**, 12349–12354.
[32] W. T. Godbey, K. K. Wu and A. G. Mikos, *Proc. Natl. Acad. Sci. USA*, 1999, **96**, 5177–5181.
[33] J.-P. Behr, *Chimia*, 1997, **51**(1–2), 34–36.
[34] A. Akinc, M. Thomas, A. Klibanov and R. Langer, *J. Gene Med.* 2005, **7**(5), 657–663.
[35] A. M. Funhoff, C. F. van Nostrum, G. A. Koning, N. M. E. Schuurmans-Nieuwenbroek, D. J. A. Crommelin and W. E. Hennink, *Biomacromolecules*, 2004, **5**(1), 32–39.
[36] P. Dubruel, B. Christiaens, M. Rosseneu, J. Vandekerckhove, J. Grooten, V. Goossens and E. Schacht, *Biomacromolecules*, 2004, **5**(2), 379–388.
[37] H. Pollard, J.-S. Remy, G. Loussouarn, S. Demolombe, J.-P. Behr and D. Escande, *J. Biol. Chem.*, 1998, **273**, 7507–7511.
[38] T. J. Deming, *Nature*, 1997, **390**(6658), 386–389.
[39] T. J. Deming, *Adv. Drug Del. Rev.*, 2002, **54**(8), 1145–1155.
[40] I. Dimitrov and H. Schlaad, *Chem. Comm.*, 2003, (23), 2944–2945.

[41] W. Zauner, M. Ogris and E. Wagner, *Adv. Drug Del. Rev.*, 1998, **30**(1–3), 97–113.
[42] M. Ohsaki, T. Okuda, A. Wada, T. Hirayama, T. Niidome and H. Aoyagi, *Bioconj. Chem.*, 2002, **13**(3), 510–517.
[43] U. Lungwitz, M. Breunig, T. Blunk and A. Gopferich, *Eur. J. Pharm. Biopharm.*, 2005, **60**, 247–266.
[44] J. Y. Cherng, P. van de Wetering, H. Talsma, D. J. A. Crommelin and W. E. Hennink, *Pharmaceutical Res.*, 1996, **13**(7), 1038–1042.
[45] P. Dubruel, B. Christiaens, B. Vanloo, K. Bracke, M. Rosseneu, J. Vandekerckhove and E. Schacht, *Eur. J. Pharm. Sci.*, 2003, **18**, 211–220.
[46] A. M. Funhoff, C. F. van Nostrum, M. C. Lok, M. M. Fretz, D. J. A. Crommelin and W. E. Hennink, *Bioconj. Chem.*, 2004, **16**(6), 1212–1220.
[47] P. Ferruti, M. A. Marchisio and R. Duncan, *Macromol. Rapid Comm.*, 2002, **23**(5/6), 332–355.
[48] S. Asayama, T. Sekine, A. Hamaya, H. Kawakami and S. Nagaoka, *Polym. Adv. Technol.*, 2005, **16**(7), 567–570.
[49] P. Dubruel, L. Dekie and E. Schacht, *Biomacromolecules*, 2003, **4**, 1168–1176.
[50] A. V. Kabanov, S. V. Vinogradov, Y. G. Suzdaltseva and V. Y. Alakhov, *Bioconj. Chem.*, 1995, **6**(6), 639–643.
[51] Y. Dong, A. I. Skoultchi and J. W. Pollard, *Nucleic Acids Res.*, 1993, **21**, 771–772.
[52] S. Stolnik, L. Illum and S. S. Davis, *Adv. Drug Del. Rev.*, 1995, **16**(2–3), 195–214.
[53] H. Otsuka, Y. Nagasaki and K. Kataoka, *Adv. Drug Del. Rev.*, 2003, **55**(3), 403–419.
[54] R. B. Greenwald, Y. H. Choe, J. McGuire and C. D. Conover, *Adv. Drug Del. Rev.*, 2003, **55**, 217–250.
[55] M. Ogris, S. Brunner, S. Schüller, R. Kircheis and E. Wagner, *Gene Ther.*, 1999, **5**, 595–605.
[56] Y. Kakizawa and K. Kataoka, *Adv. Drug Del. Rev.*, 2002, **54**, 203–222.
[57] H. K. Nguyen, P. Lemieux, S. V. Vinogradov, C. L. Gebhart, N. Guerin, G. Paradis, T. K. Bronich, V. Y. Alakhov and A. V. Kabanov, *Gene Ther.*, 2000, **7**, 126–138.
[58] C.-H. Ahn, S. Y. Chae, Y. H. Bae and S. W. Kim, *J. Control. Release*, 2002, **80**(1–3), 273–282.
[59] H. Petersen, P. M. Fechner, D. Fischer and T. Kissel, *Macromolecules*, 2002, **35**, 6867–6874.
[60] G. P. Tang, J. M. Zeng, S. J. Gao, Y. X. Ma, L. Shi, Y. Li, H. P. Too, *Biomaterials*, 2003, **24**, 2351–2362.
[61] J. W. Hong, J. H. Park, K. M. Huh, H. Chung, I. C. Kwon and S. Y. Jeong, *J. Control. Release*, 2004, **99**, 167–176.
[62] Y. H. Bae, S. Fukushima, A. Harada and K. Kataoka, *Ang. Chem. Int. Ed.*, 2003, **42**, 4640–4643.
[63] C.-H. Ahn, S. Y. Chae, Y. H. Bae and S. W. Kim, *J. Control. Release*, 2004, **97**, 567–574.
[64] M. Bikram, C.-H. Ahn, S. Y. Chae, M. Lee, J. W. Yockman and S. W. Kim, *Macromolecules*, 2004, **37**, 1903–1916.
[65] S. Fukushima, K. Miyata, N. Nishiyama, N. Kanayama, Y. Yamasaki and K. Kataoka, *J. Am. Chem. Soc.*, 2005, **127**, 2810–2811.
[66] H. Schlaad and M. Antonietti, *Eur. Phy. J. E – Soft Matter*, 2003, **10**(1), 17–23.
[67] T.-i. Kim, H. J. Seo, J. S. Choi, H.-S. Jang, J.-u. Baek, K. Kim and J.-S. Park, *Biomacromolecules*, 2004, **5**(6), 2487–2492.
[68] A. Rungsardthong, M. Deshpande, L. Bailey, M. Vamvakaki, S. P. Armes, M. C. Garnett and S. Stolnik, *J. Control. Release*, 2001, **73**, 359–380.
[69] D. Wakebayashi, N. Nishiyama, K. Itaka, K. Miyata, Y. Yamasaki, A. Harada, H. Koyama, Y. Nagasaki and K. Kataoka, *Biomacromolecules*, 2004, **5**(6), 2128–2136.
[70] T. Merdan, K. Kunath, H. Petersen, U. Bakowsky, K. H. Voigt, J. Kopecek and T. Kissel, *Bioconj. Chem.*, 2005, ASAP.

[71] S. Katayose and K. Kataoka, *J. Pharm. Sci.*, 1998, **87**(2), 161–163.
[72] E. M. Kim, H.-J. Jeong, I. K. Park, C. S. Cho, H.-S. Bom and C.-G. Kim, *Nucl. Med. Biol.*, 2004, **31**, 781–784.
[73] S. Mishra, P. Webster and M. E. Davis, *Eur. J. Cell Biol.*, 2004, **83**(3), 97–111.
[74] V. S. Trubetskoy, S. C. Wong, V. Subbotin, V. G. Budker, A. Loomis, J. E. Hagstrom and J. A. Wolff, *Gene Ther.*, 2003, **10**, 261–271.
[75] J. K. W. Lam, Y. Ma, S. P. Armes, A. L. Lewis, T. Baldwin and S. Stolnik, *J. Control. Release*, 2004, **100**, 293–312.
[76] R. C. Carlisle, T. Etrych, S. S. Briggs, J. A. Preece and K. Ulbrich, *J. Gene Med.*, 2004, **6**, 337–344.
[77] M. Turk, S. Dincer, I. G. Yulug and E. Piskin, *J. Control. Release*, 2004, **96**, 325–340.
[78] A. I. Belenkov, V. Y. Alakhov, A. V. Kabanov, S. V. Vinogradov, L. C. Panasci, B. P. Monia and T. Y. K. Chow, *Gene Ther.*, 2004, **11**(22), 1665–1672.
[79] L. Bromberg, S. Deshmukh, M. Temchenko, L. Iourtchenko, V. Alakhov, C. Alvarez-Lorenzo, R. Barreiro-Iglesias, A. Concheiro and T. A. Hatton, *Bioconj. Chem.*, 2005, **16**, 626–633.
[80] R. Kircheis, L. Wightman, A. Schreiber, B. Robitza, V. Rossler, M. Kursa and E. Wagner, *Gene Ther.*, 2001, **8**, 28–40.
[81] S. Rhaese, H. von Briesen, H. Rubsamen-Waigmann, J. Kreuter and K. Langer, *J. Control. Release*, 2003, **92**, 199–208.
[82] Riess, G. *Prog. Polym. Sci.*, 2003, **28**, 1107–1170.
[83] J. Zhu, A. Tang, L. P. Law, M. Feng, K. M. Ho, D. K. L. Lee, F. W. Harris and P. Li, *Bioconj. Chem.*, 2005, **16**, 139–146.
[84] J. H. Jeong and T. G. Park, *J. Control. Release*, 2002, **82**, 159–166.
[85] M. Bivas-Benita, S. Romeijn, H. E. Junginger and G. Borchard, *Eur. J. Pharm. Biopharm.*, 2004, **58**, 1–6.
[86] I.-S. Kim, S.-K. Lee, Y.-M. Park, Y.-B. Lee, S.-C. Shin, K. C. Lee and I.-J. Oh, *Int. J. Pharm.*, 2005, **298**(1), 255–262.
[87] M. M. Ow Sullivan, J. J. Green and T. M. Przybycien, *Gene Ther.*, 2003, **10**, 1882–1890.
[88] F. Scherer, M. Anton, U. Schillinger, J. Henke, C. Bergemann, A. Kruger, B. Gansbacher and C. Plank, *Gene Ther.*, 2002, **9**, 102–109.
[89] Y. Liu, D.-C. Wu, W.-D. Zhang, X. Jiang, C.-B. He, T. S. Chung, S. H. Goh and K. W. Leong, *Angewandte Chemie International Edition*, 2005, **44** (30), 4782–4785.
[90] R. Cartier and R. Reszka, *Gene Ther.*, 2002, **9**, 157–167.
[91] K. C. Cho, S. H. Kim, J. H. Jeong and T. G. Park, *Macromol. Biosci.*, 2005, **5**(6), 512–519.
[92] M. Licciardi, Y. Tang, N. C. Billingham, S. P. Armes and A. L. Lewis, *Biomacromolecules*, 2005, **6**, 1085–1096.
[93] P. R. Dash, M. L. Read, K. D. Fisher, K. A. Howard and M. Wolfert, *et al.*, *J. Biol. Chem.*, 2000, **275**, 3793.
[94] P. Erbacher, J.-S. Remy and J. Behr, *Gene Ther.*, 1999, **6**(1), 138–145.
[95] P. Erbacher, A. C. Roche, M. Monsigny and P. Midoux, *Bioconj. Chem.*, 1995, **6**(4), 401–410.
[96] M.-A. Zanta, O. Boussif, A. Adib and J.-P. Behr, *Bioconj. Chem.*, 1997, **8**(6), 839–844.
[97] S. S. Diebold, M. Kursa, E. Wagner, M. Cotten and M. Zenke, *J. Biol. Chem.*, 1999, **274**, 19087–19094.
[98] K. Wada, H. Arima, T. Tsutsumi, Y. Chihara, K. Hattori, F. Hirayama and K. Uekama, *J. Control. Release*, 2005, **104**, 397–413.
[99] S. R. Schwarze, K. A. Hruska and S. F. Dowdy, *Trends Cell Biol.*, 2000, **10**, 290–295.
[100] M. Zhao and R. Weissleder, *Med. Res. Rev.*, 2004, **24**(1), 1–12.
[101] B. Gupta, T. S. Levchenko and V. P. Torchilin, *Adv. Drug Del. Rev.*, 2005, **57**, 637–651.
[102] C. Rudolph, C. Plank, J. Lausier, U. Schillinger, R. H. Müller and J. Rosenecker, *J. Biol. Chem.*, 2003, **278**(13), 11411–11418.

[103] C. Plank, B. Oberhauser, K. Mechtler, C. Koch and E. Wagner, *J. Biol. Chem.*, 1994, **269**(17), 12918–12924.
[104] H.-A. Klok, *J. Polym. Sci. A, Polym. Chem.*, 2005, **43**(1), 1–17.
[105] P. van de Wetering, J. Y. Cherng, H. Talsma and W. E. Hennink, *J. Control. Release*, 1997, **49**(1), 59–69.
[106] W. T. Godbey, K. K. Wu and A. G. Mikos, *J. Biomed. Mater. Res.*, 1999, **45**(3), 268–275.
[107] P. van de Wetering, E. E. Moret, N. M. E. Schuurmans-Nieuwenbroek, M. J. van Steenbergen and W. E. Hennink, *Bioconj. Chem.*, 1999, **10**(4), 589–597.
[108] N. A. Jones, I. R. C. Hill, S. Stolnik, F. Bignotti, S. S. Davis and M. C. Garnett, *Biochimica et Biophysica Acta*, 2000, **1517**, 1–18.
[109] M. C. Deshpande, M. C. Davies, M. C. Garnett, P. M. Williams, D. Armitage, L. Bailey, M. Vamvakaki, S. P. Armes and S. Stolnik, *J. Control. Release*, 2004, **97**, 143–156.
[110] Journal of the American Chemical Society is a title from ACS publications (http://pubs.acs.org/).
[111] Advanced Materials is a title from Wiley Interscience (http://www3.interscience.wiley.com/).
[112] Bioconjugate Chemistry is a title from ACS publications (http://pubs.acs.org/).
[113] A. Akinc, D. M. Lynn, D. G. Anderson and R. Langer, *J. Am. Chem. Soc.*, 2003, **125**, 5316–5323.

Neurogenetic Imaging

Merle Fairhurst

Pain Imaging Neuroscience Group, University of Oxford, Department of Human Anatomy and Genetics, South Parks Road, Oxford OX1 3QX, UK

Abstract

From schizophrenia and Alzheimer's to pain and language, functional MRI (FMRI) and other neuroimaging techniques are lighting up the dark world of genetics. With the advent of gene expression visualisation as well as simple comparisons of genetic variants and their neural structural and functional correlates, the finer details of molecular biology are coming together with the "bigger picture" of neuroimaging. Within the last two years, however, these advances and exciting cooperative efforts have created their own scare within the public domain. Concerns over "brain privacy" invasion and ethical misuse have made a new and very real public enemy out of a technique, which is fast becoming invaluable in both the clinical and academic domains. Some of the positives and most of the negatives of the technique are now being widely broadcast, usually in an exaggerated form, to the general public. It is therefore crucial that we as a scientific community catch the wave before it breaks by providing a clear picture of where we are now and where we hope to go with this technology. Gene therapy is quite possibly our most powerful clinical tool; a way of targeting the root of disease. Imaging provides us with an image of the disease and possibly the cure. Together these techniques, however, have the scope of extending beyond research and therapeutic application. It therefore seems wise to consider and address the fears of neuro-genetic screening of physiological as well as emotional and cognitive states in healthy individuals. Although we are still several years away from the frightening reality of profiling and "mind reading" described by the press, let us take advantage of the time available to prepare the public in an informed way so that the decision of how and when the technique is used can be shared.

Keywords: FMRI, imaging, molecular, pain, privacy.

Contents

1. Introduction 78
2. Background 78
   2.1. Functional magnetic resonance imaging 78
   2.2. Neurogenetic imaging 79
3. Clinical applications 80
   3.1. Neurodegenerative diseases 80
   3.2. Pain 81
   3.3. Psychological and behavioural disorders 83
4. Ethical considerations and social implications 84
   4.1. Brain privacy 84
   4.2. Interventions 85
   4.3. A solution 85
5. Conclusion 86
References 87

1. INTRODUCTION

In the following paragraphs, neurogenetic imaging in its current and future forms will be discussed with three major areas of focus within neuroscience. Neurodegenerative diseases, pain and behavioural/psychological disorders will be used to exemplify current and possible applications of the technique. Ideas such as clinical diagnostics, molecular/pharmacological therapeutics and the role of functional magnetic resonance imaging (FMRI) can and does play within a clinical framework will be put forward. This does not exclude the role played by other forms of imaging, such as positron emission tomography (PET), already very much present within the field, but instead explores an efficient, emerging alternative technique. One should also bear in mind that medical imaging extends beyond the field of neuroscience and that it is vital that similar considerations should be made for application within others fields. The discussion will then move towards the ethical and social implications of the technique within the public domain, taking into consideration how and why the technique should be used and who should decide these limitations of acceptable use.

2. BACKGROUND

2.1. Functional magnetic resonance imaging

FMRI is a relatively new technique but it has already provided many helpful insights into the anatomy, mapping and function of many components of the central nervous system (CNS) as well as the body at large. For the purpose of this discussion, however, the focus will be on the use of and advances in FMRI of the human brain. As opposed to other imaging techniques, constant improvements of both acquisition and data analysis mean that FMRI not only allows for repeatable, non-invasive experimentation, but can also provide images with higher spatial resolution.

The technique measures contrast between oxy- and deoxy-haemoglobin content in the form of a ratio, which changes during activation of brain areas. While oxygenated blood, which is found in higher levels during activation, is diamagnetic, deoxygenated blood is paramagnetic. Placed within a magnetic field, the head produces a signal measured in terms of the ratio of oxy- to deoxy-haemoglobin. The flow of blood to activated areas produces a so-called BOLD (blood oxygenation level dependent) response which, in brief, is a change in magnetic resonance signal within this region of the image that reflects a change in the ratio of oxy- to deoxy-haemoglobin due to the increased flow of oxygenated blood to this active region. Further details can be found at: http://www.fmrib.ox.ac.uk/fmri_intro/physiology.html.

FMRI has several advantages that extend beyond its non-invasive nature and resolution benefits. Within the field of neuroimaging, task-associated activation not only allows for a global view of brain activity but does so without the need for injections of radioactive isotopes and can be performed in a relatively short total scan time. These characteristics have meant that FMRI has been introduced and has become an invaluable technique and tool within both the research and the clinical domain.

In the interest of time, discussed applications of FMRI will be restricted to the study and treatment of neurological disorders and disease. Within this area of focus, the use of FMRI is far reaching and encompasses neuropathologies such as Alzheimer's and Parkinson's as well as neuropsychological disorders such as schizophrenia and depression. As apposed to more traditional measures of "at risk" individuals, FMRI serves as a highly sensitive and reliable diagnostic device. Within sufferers, neuroimaging also enables clinicians to monitor subtle changes, even in the earliest stages of disease, which also allows for assessment of therapeutic effectiveness following treatments.

2.2. Neurogenetic imaging

Until recently neurogenetic imaging, a tool used to pioneer molecular therapeutics and research, has come to describe the comparison of structural and functional imaging of individuals with known genotypes. In the April issue of *Nature Medicine*, a technique allowing for *in vivo* detection of gene expression within mice has been reported. Ahrens *et al.* [1] triggered living cells to produce their own contrast agent (ferritin), which acts like a nanomagnet and a potent MRI reporter. This advance in itself will possibly allow for adaptation to many different tissue types and will enable pre-clinical testing of gene therapeutics and assessment of dosage protocols within transgenic animals. This opens up the field of molecular therapeutics allowing for far more accurately targeted and immediate assessment of trial pharmacological and genetic therapies within animals and it is only a matter of time before a human equivalent of the protocol is created.

The efficacy of this advanced form of imaging is highly dependent on targeting-specific biological markers that correctly identify the disease in question. These markers will include (1) cells which reflect a variety of disease characteristics, an element of the disease process itself, (2) an intermediate stage between exposure and disease onset or (3) an independent factor associated with the disease state but not causative of pathogenesis. Once found, the biomarker would be capable of not only identifying risk of disease development (antecedent biomarkers) but also aids in diagnosis and prediction of disease course (diagnostic biomarkers) and assessment of response to therapy (prognostic biomarkers). There is also the obvious need for advanced nanotechnology allowing for introduction of these contrast agents in a cell-specific manner but these concepts will not be considered within the framework of this discussion.

In the following paragraphs, three central areas of neuroscience research, namely neurodegenerative diseases, pain and behavioural/cognitive disorders, will be considered. Each will be discussed in terms of the current use of basic FMRI as a diagnostic tool and future scope for neurogenetic imaging.

3. CLINICAL APPLICATIONS

3.1. Neurodegenerative diseases

Alzheimer's disease (AD) and Parkinson's disease (PD) are progressive neurodegenerative diseases, which are now known to be present in a non-symptomatic form long before current diagnosis can be performed. Evidence of these pathological processes, as in many multi-factorial diseases, appears to vary a great deal among individuals. Early detection of clear biological markers however would allow for non-symptomatic diagnosis and, as a result, more effective treatment and possibly prevention. In the case of AD, the most common biological change is cell death. Measuring change in brain volume or density however can be misleading as cell death is associated with age as well as but not necessarily with AD. This means that distinction between aging adults with mild cognitive impairment (MCI) and those with progressive AD is not clear. Other forms of identification of "at risk" individuals currently depends primarily on family history and laboratory and genetic testing. These more traditional approaches lack the sensitivity to detect early stages of these neurological diseases. Structural and functional neuroimaging have recently been proposed as biomarkers for early detection. By using FMRI, challenging and monitoring task-associated activity in neural systems that undergo AD-induced pathological changes (medial temporal lobe, posterior cingulate), one may potentially create a clinical diagnostic screening method.

Currently, there are several alternatives of therapeutic interventions (pharmaceutical, surgical and physiological) available for the management of the motor and cognitive symptoms associated with PD. Improvements of these, however, have been slow due to the insensitive nature of existing clinical outcome measures. Progress however is being made with the use of FMRI, where for example Elsinger *et al.* [2] have measured partial normalisation of brain activation patterns following a finger-tapping task, comparing patients with and without dopamine replacement therapy. These therapeutic differences, by contrast, had not been observed following traditional behavioural testing procedures. This highlights how more sensitive therapeutic monitoring allows for fine-tuning of existent therapies and possibly directs research towards more efficient treatment options.

By identifying risk genes, for example, possible therapeutic targets are highlighted. In the case of AD, four main genes have been linked to different forms of the disease including missense mutations in presenilin 1 (PS1), mutations of presenilin 2 (PS2) and in a few rare familial AD cases, mutations in and over-expression of amyloid precursor protein (APP) surrounding the β- and γ-secretase sites. These mutations all result in increased A,42 production but as these mutations and early onset AD are rare, there is still no commonly identified deterministic gene for the sporadic form of AD. "At risk" individuals lacking in ε4 alleles appear to have the highest amounts of CSF (cerebrospinal fulid) Aβ followed by those individuals with one allele and then by those with two. A preliminary study suggests that a dose-dependent effect of ε4 on the amount of Aβ1-42 in CSF could be used to track changes of the level of CSF Aβ, a decrease which might suggest the build up of plaques forming in the brain. This sort of test could be used diagnostically in cognitively normal individuals with family history of AD but the utility of ε4 alone as a definitive, predictive biomarker is currently limited. It may have value in conjunction with other measurements including correlating known genotype with changes in FMRI neuronal activation, observed by Bookheimer *et al.* [3] to have some advantage of assessing risk before symptomatic memory decline.

Studies continue to search for other useful genetic indicators of sporadic AD and, in particular, much work is being devoted to the development of non-toxic contrast agents targeting proteins known to differentiate healthy individuals from those with symptomatic and, ideally, asymptomatic AD. Currently under investigation is putrascine-gadolinium-Aβ (PUT-Gd-Aβ), which if preliminary evidence holds true, has the ability to identify plaques within mice both *in vivo* and post mortem [13].

Research into therapeutic use of transgenes in PD has also reached a stage where much benefit could be derived from neurogenetic imaging. Targets of interest include Parkin and α-synuclein, the latter of which seems to be of particular importance as, at least in some early-onset familial cases of PD, a "single gene defect, which alone seems to be sufficient to determine PD phenotype" [4]. The phenotype in this case is the presence of Lewy bodies, intracytoplasmic inclusion bodies, which are a specific pathological feature of PD.

3.2. Pain

The network underlying the perception of pain combines nociceptive, emotional and cognitive input. The introduction of these varying inputs allows, for instances, where the relationship between stimulus intensity and unpleasantness is no longer a linear one [5]. Instead, owing to modulation by the various circuits within the network, perception of the stimulus can either be dulled or heightened by factors such as emotion (past experiences or present feelings), attention and anticipation. It is this ability of the brain to adapt our perception of pain that is of interest in a clinical sense. A better understanding of the machinery behind these modulatory pathways might lead to treatments that make better use of this endogenous analgesic system. Cognitive behavioural therapy, as a treatment of chronic pain is one example of the use of the anti-nociceptive pathway.

Both the ascending and descending modulating loops of the pain network are being studied using varied methods of experimentation. Neuroimaging has proved especially helpful in identifying and studying the CNS structures involved. Thus far, animal studies have greatly advanced our understanding of this wiring and it is the emergence and development of PET and FMRI that these studies are being re-evaluated in humans.

FMRI studies of pain have focused primarily on exploring the structures within and that work in conjunction with the so-called pain matrix. This fairly new field of research however, is now delving deeper into the details of the associated pain pathways, including how the components interact to form an integrated whole. This may allow for more specific targeting of pain-related disease and pain mechanisms in the future. Presented below are two examples of how individual pain sensitivity is being explored so as to better understand pain perception as a whole and, based on this knowledge, possible analgesic interventions.

Genes responsible for individual variation in sensitivity to and inhibition of pain, via modulatory pathways, could be invaluable in clinical medicine. Zubieta *et al.* [6] has been particularly successful in isolating what has come to be known as the "pain tolerance gene" – COMT gene. In this study, the group used the more common form of neurogenetic imaging, described in previous paragraphs, where individuals of different known genotypes, in this case variances in the met/met val/val polymorphisms, were compared using their structural and functional neural correlates.

On a different track, groups have focused on the placebo response, which is thought to be an important contributor to the efficacy of analgesic interventions. There are however, many unanswered questions with regard to the mechanisms by which the response produces its effects. The placebo response has been associated with the descending modulatory pathways and placebo-induced pain reduction is known to be a rewarding event. Many studies involving gambling and other reward protocols have tried to link an associated genotype to individual variances in reward processing. During a 2003 pilot study in humans, Compton *et al.* [7] proposed that variances in the human micro-opioid receptor gene (OPRM) may be associated with not only pain tolerance but also opiate addiction. A better understanding of polymorphisms of this particular gene, aided by neurogenetic imaging, exemplifies how very distinct benefits and disadvantages exist for each piece of information. In this case, a more firm understanding of placebo and reward systems would advance our attempts at utilising endogenous analgesic systems for pain relief as well as possibly lead towards a cure for opiate and other substance abuse patients. On the other hand, information pertaining to an individual's susceptibility for substance abuse, obtained from neurogenetic screening for OPRM identifying a similar gene or a combination of genes, could alter the way that individual is received by society. Other examples and issues surrounding the ethics of neurogenetic imaging will follow in a separate and subsequent section.

3.3. Psychological and behavioural disorders

The applicability and use of FMRI within the realm of psychological and behavioural disorders is considerably great. In contrast with both pain and the neurodegenerative diseases described in the preceding paragraphs, our understanding of the physiological and biological basis of these disorders is far more basic. This however can be seen as an advantage as the current and potential role played by FMRI to explore these disorders is all the greater [12].

In the case of schizophrenia, for example, this technique has already shed much light on the effects of pharmacological treatment on neurocognitive correlates of the disorder's neuroanatomic basis. The development of cognitive tasks that assess working memory function with FMRI have led to the consistent finding of dorsolateral prefrontal cortex (DLPFC) as being one of, if not the locus of dysfunction in the patho-physiology of, schizophrenia. Pathological changes within the region, known to be critical for working memory, result in the basic symptomology of the disorder, namely basic deficits in strategic processing including inadequate filtering of irrelevant stimuli or maintenance of a goal-directed line of thought.

Beyond FMRI's ability to diagnose, we are also fortunate to be seeing much progress in pharmacological treatments of schizophrenia due to the technique. With it one is able to measure the extent to which a drug reaches its targeted neural system thereby providing a more effective clinical outcome measure. By comparing working memory associated activity, determining differential efficacy of treatment alternatives, discriminating between drug effects and placebo as well as testing the efficacy of surgical procedures is made possible.

Attention deficit hyperactivity disorder (ADHD) is a common childhood neurodevelopmental, behavioural disorder which seems to persist in some cases into adolescence and sometimes even into adulthood. Characteristic cognitive impairment includes deficits in response inhibition, working memory, sustained attention, timing perception and reproduction and conceptual reasoning. Diagnosis until recently has been limited to behavioural and psychological assessment monitoring, for example, poor academic achievement or delinquency.

Until the advent of neuroimaging, our knowledge of the brain networks mediating cognitive and behavioural dysfunction has been limited and thus, treatment of ADHD has been restricted to stimulants such as methylphenidate

(MP), observed to ameliorate a broad range of the characteristic deficits. Using similar working memory tasks and monitoring hypoactivation of a right inferior frontoparietal network [8], known to be associated with attentional mechanisms, impaired performance in off medication ADHD patients has been observed. Compared to healthy controls, children with ADHD had less striatal activation during a cognitive-inhibition task [9]. MP increased striatal activation in patients with ADHD but decreased striatal activation in controls.

The COMT gene known to play a crucial role in the regulation of central dopaminergic systems, and described in a previous paragraph in relation to pain tolerance, has been observed in individuals with the val/val genotype to allow for more efficient conflict resolution using the stroop conflict test [10]. The valine allele of COMT, producing relatively higher levels of enzyme activity and therefore lower relative amounts of extrasynaptic dopamine, has been examined using neurogenetic imaging studies and has been correlated with lower activity of the DLPFC [11] and associated with schizophrenia. Should a cost effective, clinical screening process similar to this become available, it would allow for an invaluable diagnostic tool in a field of medicine which has suffered greatly from very little in the way of a concrete biological basis.

4. ETHICAL CONSIDERATIONS AND SOCIAL IMPLICATIONS

Beyond its clinical use in the advancement of molecular/pharmacological therapeutics (gene therapy included) and diagnosis, neurogenetic imaging may very well enter the public domain. The question as to the purpose of such a technique beyond the current and future clinical roles described in the preceding section is an important one.

The possible uses of neurogenetic imaging can be divided into two clear categories namely (1) physiological/behavioural/cognitive screening of healthy individuals and (2) interventions which alter brain activity. Of particular concern is the maintenance of privacy of the information obtained from the scans and, with the information from the scanning procedure, who and how one decides what subsequent interventions are seen as acceptable.

4.1. Brain privacy

Current media hype has focused primarily on the idea of brain privacy. In a way quite unique to this medical tool, brain imaging produces a very tangible, if indirect, readout of an individual. The ability to describe an individual's behavioural tendencies, for example, is far more frightening and real a concept than a particular gene susceptibility for those outside of the scientific domain. And once this individual information is available, either on one of the various FMRI databases already being created or alternatively as part of a personal record, who should be able to peruse/use it? Should this screening facility be provided by someone other than a medical practitioner, the concept of "patient" confidentiality would need to be revised. Concerns have been raised over how this information could affect society's perception not only of an individual but also the importance of individuality. Behavioural screening, for example, indicating a propensity for aggression or presence of a behavioural/cognitive disorder such as ADHD, may have untold implications on how that person is medicated and more importantly how they are received into society, treated professionally, or in the case of children how he/she is educated. The social ramifications of trait screening need to be evaluated, as they will very much determine subsequent action and interventions.

4.2. Interventions

Fears of cognitive enhancement or "designer babies" are merely two examples borne out of reports of possible extended applications of FMRI screening. How realistic these types of interventions are is debatable but the clinical interventions alluded to in the preceding clinical discussion do bring into question or should at least prompt us to consider what is deemed acceptable use of molecular therapeutics. In the case of the discussed COMT gene, polymorphisms have been observed to lead to schizophrenia, ADHD and/or low pain sensitivity within certain individuals. Molecular alterations within individuals baring these "undesirable" genotypes may seem clinically and ethically sound and, in theory, adhere to our motivating principle of improving quality of life. Is it possible though to want to "solve" or "treat" too much? Should, for example, limitations be set in place to restrict fetal FMRI screening or are pre-natal diagnostics essential?

4.3. A solution

The time available, before the science catches up with the science fiction created by media hype and public imagination, is a crucially important window and equally a potential chance for scientists to repair our somewhat tainted reputation. As described and exemplified above, neurogenetic imaging is a tool with the capacity to be a powerful medical tool but may serve as an equally powerful money-making diagnostic tool within the so-called "worried healthy" and whose resulting data may be misused to divide men and women based on desired and undesired physical and mental traits. A corrected view of the scientific community will not only enable further advances to be made with the support of the people for which the advances are destined, but also allows for more value to be placed in our opinions, when it comes to deciding appropriate use of the tool our professional opinions are needed. In the case of pure gene therapy as well as the continually contentious and general case of animal testing, we have witnessed how public ill-will can limit not only the science in itself but also rational, emotionally detached discussion surrounding the subject at hand. So how do we avoid a similar stigma for neurogenetic imaging from being created? The process would seem to involve finding answers to two important questions.

What can realistically be derived from neurogenetic imaging?

Scientific experimentation, through a process of inductive reasoning, provides a better idea of the bigger picture. As members of the scientific community, we understand and take this fact for granted and in everything we report we almost always humbly accept that our findings are merely a piece of the puzzle from which no absolute conclusion can be drawn. It is within the media's interest however to be able to narrow down and simplify information into a single headline. The current format of information communication is quite clearly unsuitable for transmitting the complexities and subtleties of neurogenetic imaging. It would therefore be in everyone's best interest to determine first, what information is necessary to provide to give an adequate description of the technology and its possible applications and second, who should and how this information should be conveyed to the public. This would hopefully limit fearful and exaggerated versions of the reality as well as lead us to answering the second of the two questions.

What do we want from neurogenetic imaging?

A vital element to keep in the forefront of this discussion and indeed something to stress when attempting to win back the public's favour is making known our general intentions. Particularly within the medical sciences, research and application of findings are at least in some part motivated by a desire to improve the quality of life. These good intentions however do not always stretch outside of the clinical sphere in and for which the tool was designed. It would seem an important step to remind others of, and keep in mind ourselves, our own desired outcomes for this tool.

5. CONCLUSION

Answers to these and many new questions, a list of which is constantly being added to by a growing number of bioethic groups, will need to be addressed. This will need to be achieved, if possible, before a negative public view limits the true scope for very real and invaluable clinical benefits. It seems essential that as a primary step we, as a scientific community, regain trust in the eyes of the public. A possible means of communication is through the media, which has the potential to link lay and scientific groups, elucidating the real scope of this technology and initiating dialogue between all necessary parties.

REFERENCES

[1] E. T. Ahrens, U. DeMarco, G. Genove, H. Xu and W. F. Goins, A new transgene reporter for magnetic resonance imaging, *Nat. Med.*, 2005, **11**(4), 450–454.

[2] C. L. Elsinger, S. M. Rao, J. L. Zimbelman, N. C. Reynolds, K. A. Blindauer and R. G. Hoffmann, Neural basis for impaired time reproduction in Parkinson's disease: An FMRI study, *J. Int. Neuropsychol. Soc.*, 2003, **9**(7), 1088–1098.

[3] S. Y. Bookheimer, M. H. Strojwas, M. S. Cohen, A. M. Saunders, M. A. Pericak-Vance, J. C. Mazziotta and G. W. Small, Patterns of brain activation in people at risk for Alzheimer's disease, *N. Engl. J. Med.*, 2000, **343**(7), 450–456.

[4] M. H. Polymeropoulos, C. Lavedan, E. Leroy, S. E. Ide, A. Dehejia, A. Dutra, B. Pike, H. Root, J. Rubenstein, R. Boyer, E. S. Stenroos, S. Chandrasekharappa, A. Athanassiadou, T. Papapetropoulos, W. G. Johnson, A. M. Lazzarini, R. C. Duvoisin, G. Di Iorio, L. I. Golbe and R. L. Nussbaum, Mutation in the alpha-synuclein gene identified in families with Parkinson's disease, *Science*, 1997, **276**(5321), 2045–2047.

[5] P. Petrovic, K. Carlsson, K. M. Peterson, P. Hansson and M. Ingvar, Context-dependent amygdala deactivation during pain, *Neuroimage*, 2001,**13**, S457.

[6] J. K. Zubieta, M. M. Heitzeg, Y. R. Smith, J. A. Bueller, K. Xu, Y. Xu, R. A. Koeppe, C. S. Stohler and D. Goldman, COMT val158met genotype affects mu-opioid neurotransmitter responses to a pain stressor, *Science*, 2003, **299**(5610), 1240–1243.

[7] P. Compton, D. H. Geschwind and M. Alarcon, Association between human mu-opioid receptor gene polymorphism, pain tolerance, and opioid addiction, *Am. J. Med. Genet. B. Neuropsychiatr. Genet.*, 2003, **121**(1), 76–82.

[8] K. Rubia, S. Overmeyer, E. Taylor, M. Brammer, S. C.Williams, A. Simmons and E. T. Bullmore, Hypofrontality in attention deficit hyperactivity disorder during higher-order motor control: A study with functional MRI, *Am. J. Psychiatry*, 1999, **156**(6), 891–896.

[9] C. J. Vaidya, G. Austin, G. Kirkorian, H. W. Ridlehuber, J. E. Desmond, G.H. Glover, J.D. Gabrieli, Selective effects of methylphenidate in attention deficit hyperactivity disorder: A functional magnetic resonance study. *Proc. Natl. Acad. Sci. USA*, 1998, **95**:14494–14499.

[10] T. Sommer, J. Fossella, J. Fan and M. I. Posner, Inhibitory control: Cognitive subfunctions, individual differences and variation in dopaminergic genes, in *The Cognitive Neuroscience of Individual Differences – New Perspectives* (eds. I. Reinvang, M. W. Greenlee, and M. Herrmann), Bibliotheks- und Informationssystem der Universitat Oldenburg, Oldenburg, Germany, 2004.

[11] M. F. Egan, T. E. Goldberg, B. S. Kolachana, J. H. Callicott, C. M. Mazzanti, R. E. Straub, D. Goldman and D. R. Weinberger, Effect of COMT Val108/158 Met genotype on frontal lobe function and risk for schizophrenia, *Proc. Natl. Acad. Sci. USA*, 2001, **98**(12), 6917–6922.

[12] A. Gropman, Imaging of neurogenetic and neurometabolic disorders of childhood, *Curr. Neurol. Neurosci. Rep.*, 2004, **4**(2), 139–146.

[13] C. R. Jack Jr., Antecedent biomarkers conference, 2003.

Implications of Fetal Gene Therapy for the Medical Profession

Thomas Meyer

Leibniz-Forschungsinstitut für Molekulare Pharmakologie, Berlin-Buch, Germany

Abstract

Gene therapy, the treatment of disease by the transfer and sustained expression of an exogenous normal gene in somatic cells of a patient, is aimed to correct some types of severe genetic disorders. For most inherited genetic diseases, this therapeutic intervention should occur as soon as possible in early life before the onset of any fetal tissue damage. The rationale of fetal gene therapy is to transform a chronic, debilitating and often fatal disease into one that can be cured before birth, rather than merely to have it managed through the course of the patient's lifetime. In the past years, the challenge of *in utero* gene transfer has evoked an extensive debate not only on the safety and practicality of this novel therapeutical approach, but also a broad range of ethical and theological concerns. These controversies have focused mainly on ethical, legal, and social issues, such as questions concerning the ethical status of the embryo as a subject for clinical trials, and concerns were raised because of the violation of the genetic integrity of the fetus. Numerous publications relate to the balancing of risks and benefits for the mother and the fetus, the safety and efficacy of the approach, and the transgenerational risks to the germ-line. The impact of fetal gene therapy for the medical profession, however, has gained much less attention, even though the prospect of gene therapy in the womb promises to change the practice of prenatal medical treatment fundamentally. The present article focuses on the implications of *in utero* gene intervention for the patient–physician relationship and asks how the decision making in the context of fetal gene therapy will be balanced between the general practitioner, other health care professionals, and the mother. Furthermore, this paper asks how the acceptance of fetal gene transfer by different sections of the society will influence the attitude and expectations of both the mother and the treating physician towards abortion.

Keywords: Fetal gene therapy, ethical consideration, physician.

Contents

1. Ethical objections to fetal gene therapy 90
2. *In utero* gene transfer as a new therapeutical option 91
3. The medical profession in the context of fetal gene therapy 92
References 94

1. ETHICAL OBJECTIONS TO FETAL GENE THERAPY

The stable integration of an exogenous DNA sequence into a patient with the aim of correcting phenotypic abnormalities has the potential to either heal or at least relieve an otherwise intractable genetic disorder. Despite some advances in gene therapy for a variety of inherited or acquired diseases, recently reported tragic setbacks have shed light on the inherent shortcomings of current gene transfer techniques. In particular, the questions of which and whether available delivery or transfer systems are acceptably safe enough, and should thus be preferred in clinical settings, remain unanswered and a matter of controversy. Although it is too early to predict the significance of somatic gene therapy for future medicine, there are many reasons for moving forward to optimise gene transfer protocols. However, much more work is needed to assess the clinical applicability, efficacy, and safety of different delivery systems such as retroviral or adeno-associated virus, liposome, and adenovirus. All of these vector systems have desirable features of their own but also obvious disadvantages, such that future investigations will have to focus on their clinical utility and potential adverse effects [1–5]. Thus, ethical issues surrounding somatic gene therapy in children or adults are primarily those concerned with safety, informed consent and the privacy and confidentiality of genetic patient data [3,6–10].

For somatic gene therapy, specially designed delivery systems are needed to ensure that the transgene and the vector are only administered locally and that the transgene is expressed exclusively in the target cells. Thus, reasonable precautions are necessary to prevent that the vector has access to the circulation and may be integrated at sites where its expression is not desired. Nevertheless, it is generally accepted that genetic manipulation in somatic cells precludes the possibility that the transgene is transmitted via genetically altered gametes to future generations. For fetal gene therapy, however, the risk of germ-line transmission may not be negligible [3]. Indeed, it is conceivable that the delivery and expression of a foreign DNA fragment in the fetus, which is achieved in mid-to-late pregnancy, carries the risk to transmit an iatrogenically introduced transgene to the offspring of the fetus itself. Given that the gene of interest is stably integrated into the genome of germ-line cells and is functionally expressed there, this will obviously affect the genetic integrity of future generations of people. Ethical objections and concerns arising from this issue have been extensively discussed in the literature and are expected to limit the acceptance of this new therapeutical approach [2,3,8,10,11,12].

For fetal gene therapy, an additional objection has been raised, as it cannot be excluded that the maternal circulation takes up significant amounts of the *in utero*-administered DNA. After being integrated into maternal cells, the expression of the foreign DNA may either be of no consequences or interfere with important cellular functions giving rise to adverse site effects.

Thus, genetic alteration of fetal tissue is potentially dangerous not only for the fetus, but also for the mother [3–5,7,8,11]. The risk that fetal gene delivery has adverse effects on the mother may indeed be very low, however because of missing data, this risk is inestimable. Generally, it is argued that the gonads are highly compartmentalised from the rest of the body thus offering effective protection against circulating DNA molecules [3]. Nevertheless, the risk of gene integration into the germ-line is difficult to assess, since valid studies on this issue have not been published yet.

Because fetal gene therapy can only be performed in conjunction with prenatal genetic diagnosis, guidelines are required which will have to respect the privacy of the genetic data obtained. Importantly, detailed protocols for *in utero* gene transfer have to be elaborated in order to prevent irreversible damage during gestation and to hopefully allow the birth of a normal baby without clinical symptoms [13].

2. *IN UTERO* GENE TRANSFER AS A NEW THERAPEUTICAL OPTION

The certainty to give birth to a child suffering from a severe genetic disorder is a heavy burden for the parents and is frequently associated with anxiety, feelings of guilt, and emotional stress. Beyond clinical considerations like therapy planning and assignment of the recurrence risk for siblings, significant and long-lasting emotional relief for the parents is urgently needed and the treating physician has to support the parents coping with such an unexpected situation. A trained and experienced physician should assist the woman in making a fully free and informed choice between different therapeutic options. The treating physician bears the full responsibility for making such a recommendation but, based on the patient's autonomy, the woman reserves the right to finally decide which treatment strategy should be followed. The professional opinion of the physician must be based solely on the medical assessment of the patient's health interest and requires an appropriate and complete consultation, so as to prevent hasty or ill-considered decisions.

Abortion, as one of the therapeutic proposals, carries serious health risks for the woman and in this situation she may decide to deliver a baby suffering from symptoms of an underlying genetic disorder. When both alternatives, either iatrogenic termination of pregnancy or therapeutic nihilism seems unbearable for the mother, fetal gene therapy is regarded as a promising causative intervention to prevent or at least protract the onset of clinical symptoms. Provided that future advances have successfully overcome the inherent technical difficulties accompanying *in utero* gene transfer, gene therapy raises some hope to allow the birth of a normal child without clinical symptoms [1,5,7].

For preimplantation genetic diagnosis, early embryos are obtained prior to implantation and are then biopsied with very fine glass needles under microscopic observation to obtain blastomeres. Following testing, only those embryos with normal genetic results are selected for immediate transfer into the uterus. In contrast to preimplantation diagnosis, which detects a genetic defect before establishing pregnancy, fetal gene therapy offers the potential to treat a fetus after implantation. Advances in ultrasound-guided percutaneous administration of therapeutics allow for safe and effective delivery of gene constructs either topically into targeted fetal tissues or systemically into the fetal circulation. An alternative approach to the direct administration of gene constructs into the fetus or the amnion fluid involves the removal of fetal cells and their genetic manipulation outside the body. The genetically modified cells are then reimplantated at the desired anatomical site within the embryo [14].

Both strategies, delivery of recombinant vectors for *in vivo* use and reimplantation of *ex vivo*-manipulated fetal cells, are going to broaden the restricted repertoire of prenatal treatment concepts. Clearly, the prospects for *in utero* gene therapy constitute a third option for families who are faced with the prenatal diagnosis of a severe genetic disorder. Given that the essentially random and non-homologous mode of gene integration into the genome can be effectively prevented and that all other problems associated with the safety of this obstetric intervention are satisfyingly solved, gene delivery in early fetal life may no longer be a wishful thinking. Instead, it may offer a feasible strategy to help women with the decision of whether to continue a pregnancy.

3. THE MEDICAL PROFESSION IN THE CONTEXT OF FETAL GENE THERAPY

The success of fetal gene therapy critically depends on future progress in the rapidly evolving field of genetic recombination technology and vector design. When experimental studies have established the proof of principle that fetal gene transfer can achieve therapeutic effects, however many questions will still remain: What are the risks to the fetus for accidental damage and for gene transfer to the germ-line? And what is the risk for the mother? What exact advantages offers *in utero* gene delivery over postnatal therapies and in which cases should fetal treatment be preferred? What will be the quality of life for the treated child and what long-term results may be expected? Other unsolved questions in this context are: How is informed consent for this new approach to be obtained and what is the legal and ethical status of the fetus? Who will have access to this new therapy and who will bear the economic costs? Will the costs be affordable for the society? For most, if not all, of these topics there must be a clear answer and reliable and valid data must guide the decision as to whether gene delivery in the womb is considered a viable therapeutic alternative [8].

The treating physician, who faces both the expectations and the uncertainty associated with this emerging therapeutic option, bears responsibility for the pregnant woman and her choice of the optimal treatment strategy. An appropriate and comprehensive medical examination of the patient is mandatory and the physician has to consider the extreme emotional stress and ethical burden for the pregnant woman. The patient–physician relationship in this situation is unique in that, more so than for other medical fields, it is characterised by uncertainties with regards to benefits and risks of different treatment options. These uncertainities will obviously hamper decision making.

Of course, all recommended medical interventions and even those that had been omitted must conform to the doctrine of informed consent, in that the pregnant woman holds the ultimate decision on which medical procedure should be followed. Intrauterine delivery of gene vehicles for fetal targeting is only one putative approach in this situation, and the physician is well advised to concentrate on more established therapies whose outcome is currently more predictable. Despite its conceptual appeal, protocols for fetal gene therapy will have to be embedded into the broad range of already existing treatment strategies. Moreover, for every monogenic disease entity or even more for a defined mutation or polymorphism of a certain gene, separate cost–benefit analyses will have to be performed on an individual base. These considerations as well as analysis of the genetic and somatic constitution of the affected embryo have to be intensively discussed between physician and the mother in an atmosphere grounded in trust rather than merely informed consent.

An important issue in this context is the privacy protection for genetic and clinical data. As fetal gene therapy cannot be performed without genetic testing, medical confidentiality with respect to the DNA sequencing data of the fetus must be strictly respected. As genes and mutations within them determine the genetic identity and inheritance of the fetus, these data must be handled with professional diligence. In all cases of prenatal genetic testing, the physician must sign a statement from the pregnant woman that informed consent was obtained prior to testing. The results of the genetic testing should be interpreted in the light of the expected clinical presentation. When communicating these results to the mother, the physician must assure that she fully realises the clinical consequences thereof.

The ethical legitimacy of fetal gene therapy with regard to the protection of both woman and fetus will profit greatly from individual treatment attempts which are conducted in well-planned research projects guided by scientific evaluation. Following established guidelines, current regulations governing informed consent require that all considerable medical information should be given to the pregnant woman. The disclosure of all relevant data to the woman should be more detailed than that routinely provided to patients in standard clinical care, as has been proposed for genetic research trails [6]. For the first fetal gene therapy protocols it can be expected that individual healing attempts resist the traditionally maintained categorisation as purely research or purely therapy [6]. Therapeutic uncertainties and potentially hazardous complications of fetal gene delivery require that all associated procedures be exclusively performed in centres with long-standing expertise in obstetric surgery for safe introduction of recombinant DNA technology. In the first line of clinical trails on human prenatal gene therapy it is recommended that only specialised and experienced centres should participate. In contrast to the generally held assumption that in cases of desperately ill patients the physician is allowed to offer experimental intervention, it is argued here that innovative and potentially life-supporting procedures should not be categorically banned, but instead be performed under strict regulations and accompanied by scientific programmes. This requires that in carefully designed animal experiments, the benefits with regards to the health of the mother and the fetus override by far the risk for putative complications. At this point, however, the pregnant woman, her practitioner, and the obstetrician have to accept the challenge that fetal gene therapy is still far from being applicable in routine clinical practice. Future developments in this field will show whether the diagnosis of an inherited disease and the emerging potentials for fetal treatment will change attitudes towards pregnancy and abortion.

REFERENCES

[1] E. D. Zanjani and W. F. Anderson, Prospects for *in utero* human gene therapy, *Science*, 1999, **285**, 2084–2088.

[2] C. Coutelle, M. Themis, S. Waddington, L. Gregory, M. Nivsarkar, S. Buckley, T. Cook, C. Rodeck, D. Peebles and A. David, The hopes and fears of *in utero* gene therapy for genetic disease, *Placenta*, 2003, **24**, S114–S121.

[3] K. R. Smith, Gene therapy: Theoretical and bioethical concepts, *Arch. Med. Res.*, 2003, **34**, 247–268.

[4] S. N. Waddington, N.L. Kennea, S. M. Buckley, L. G. Gregory, M. Themis and C. Coutelle, Fetal and neonatal gene therapy: Benefits and pitfalls, *Gene Ther.*, 2004, **11**, S92–S97.

[5] S. N. Waddington , M. G. Kramer, R. Hernandez-Alcoceba, S. M. Buckley, M. Themis, C. Coutelle and J. Prieto, *In utero* gene therapy: Current challenges and perspectives, *Mol. Ther.*, 2005, **11**, 661–676.

[6] L. R. Churchill, M. L. Collins, N. M. P. King, S. G. Pemberton and K. A. Wailoo, Genetic research as therapy: implications of "gene therapy" for informed consent, *J. Law Med. Ethics*, 1998, **26**, 38–47.

[7] C. Coutelle, A. M. Douar, W. H. Colledge and U. Froster, The challenge of fetal gene therapy, *Nat. Med.*, 1995, **9**, 864–866.

[8] C. Coutelle and C. Rodeck, On the scientific and ethical issues of fetal somatic gene therapy, *Gene Ther.*, 2002, **9**, 670–673.

[9] J. Spink and D. Geddes, Gene therapy progress and prospects: Bringing gene therapy into medical practice: the evolution of international ethics and the regulatory environment, *Gene Ther.*, 2004, **11**, 1611–1616.

[10] D. Sicard, Ethical questions raised by *in utero* therapeutics with stem cells and gene therapy, *Fetal Diagn. Ther.*, 2004, **19**, 124–126.

[11] J. C. Fletcher and G. Richter, Human fetal gene therapy: Moral and ethical questions, *Hum. Gene Ther.*, 1996, **7**, 1605–1614.
[12] M. D. Evans, J. Kelley and E. D. Zanjani, The ethics of gene therapy and abortion: public opinion, *Fetal Diagn. Ther.*, 2005, **20**, 223–234.
[13] A. L. Caplan and J. M. Wilson, The ethical challenges of *in utero* gene therapy, *Nat. Genet.*, 2000, **24**, 107.
[14] J. R. Meyer, Human embryonic stem cells and respect for life, *J. Med. Ethics*, 2000, **26**, 166–170.

Ethical Issues in Gene Therapy Research: An American Perspective

Rebecca S. Feinberg

Abstract

Gene therapy has been the subject of significant ethical discussion. Gene therapy has spawned great due to its potential for affecting the human genome. There are two forms of gene therapy currently being researched, somatic-cell gene therapy and germ-cell gene therapy. Somatic-cell gene therapy research seeks to alter or manipulate specific organ or tissue cells. Germ-cell gene therapy does not seek to cure a disease currently being suffered, rather it seeks to alter the individual's genome so that a disease-causing mutation will not be inherited by the individual's progeny. In this paper we will discuss the arguments in favor of gene therapy research, the arguments against gene therapy research, general societal concerns regarding gene therapy and the American approach to gene therapy research regulation.

Keywords: gene therapy, somatic-cell, germ-cell, ethics, research ethics.

Contents

1. Introduction 99
2. Arguments in favor of gene therapy research 100
3. Arguments against gene therapy research 102
4. General societal concerns regarding gene therapy research 104
5. The american approach to gene therapy regulation 106
References 107

1. INTRODUCTION

Gene therapy, like all previous areas of cutting-edge scientific research, has been the subject of significant ethical discussion. Gene therapy has arguably spawned more debate than other emerging medical breakthroughs due to its potential for affecting the human genome. There are two forms of gene therapy currently being researched, somatic-cell gene therapy and germ-cell gene therapy. Somatic-cell gene therapy research seeks to alter or manipulate specific organ or tissue cells. The idea is that these altered cells, altered to have the defective portion of the gene removed and replaced by a 'healthy' gene, can then be reintroduced into the patient to remedy the disease caused by the defective portion of the gene. In contrast, germ-cell gene therapy does not seek to cure a disease currently being suffered; rather it seeks to alter the individual's genome so that a disease-causing mutation will not be inherited by the individual's progeny. Thus, germ-cell therapy has the intent of altering the human genome for generations to come. There are subcategories of both somatic-cell and germ-cell therapy dedicated to enhancement, but these will not be discussed in this paper. In this paper, we will discuss the arguments in favor of gene therapy research, the arguments against gene therapy research, general societal concerns regarding gene therapy, and the American approach to gene therapy research regulation.

2. ARGUMENTS IN FAVOR OF GENE THERAPY RESEARCH

Proponents of gene therapy focus their endorsement on Utilitarian arguments. Four primary arguments made in favor of gene therapy research are (1) scientific freedom, (2) effect and utility, (3) potential for disease eradication, and (4) reproductive freedom. The first three arguments are grounded in Utilitarian philosophy, in other words, what actions will maximize the greatest good for the most people. The last argument is grounded in Libertarian philosophy, in other words, what actions will maximize the parents' autonomy to actualize their reproductive freedom. Each of these arguments is sequential, gaining validity from each previous argument. Although there are many additional justifications in favor of gene therapy research, we will focus on these four because they are the most common.

The advancement of all research is based upon the principle of scientific freedom. Scientific freedom is the premise that all knowledge is good and therefore scientists should be permitted and even encouraged to pursue any line of research that interests them. This is well stated by a vocal advocate of gene therapy, Dr. Zimmerman, who states: "the prevailing ethic of science and medicine is that knowledge has intrinsic value, and that its pursuit should not be impeded except under extraordinary circumstances." [1]. Consider penicillin, which was a research experiment gone awry. A drug that is the foundation of antibiotic treatment in modern medicine was a scientific mistake, made under the premise of scientific freedom. In other words, had research been restricted to pre-approved topics, discoveries that we now consider as essential elements of medical practice would never have been found. In other words, research in the area of gene therapy may do far more than lead to effective gene therapy treatments. Along the path of gene therapy research, scientists may discover something that becomes an essential and vital part of medical treatment. Thus, the Utilitarian would say that gene therapy research has potential to do good for the greater population and is therefore justified.

Effect and utility are an extension of the scientific freedom philosophy. Specifically, effect and utility rely upon the potential treatments and cures that could be derived from gene therapy research. Fletcher and Anderson argue that medical science and thereby the scientists doing research, have a moral obligation to make available the best possible treatments [2]. This argument is particularly strong when comparing somatic-cell gene therapy and germ-cell gene therapy. When germ-cell gene therapy progresses to the point at which it is commonplace medical practice, somatic-cell gene therapy will no longer be necessary [3]. Diseases which would have required treatment with somatic-cell gene therapy would have been prevented by germ-cell gene therapy performed either around the time of fertilization or in a previous generation. In other words, a child who was treated preconception or whose parents were treated with germ-cell gene therapy may never need to be treated. Thus, a child who may have suffered either the disease or its treatment will never even be aware that a potential problem may have existed.

The concept of disease eradication stems directly from the effect and utility argument. Specifically, it is a continuation of the germ-cell gene therapy versus the somatic-cell gene therapy logic. Once the germ-cell gene therapy becomes a common practice, somatic-cell gene therapy will no longer need to be utilized because the diseases that would have required the somatic-cell gene therapy will be prevented prior to requiring treatment with the somatic-cell gene therapy. According to Szebik and Glass, germ-cell gene therapy will ultimately become prophylactic [4]. This argument can be extended to encompass disease eradication as follows: Since germ-cell gene therapy, in theory, repairs the diseased portion of the parents' genome prior to conception of the child, the child will not suffer from the disease. Not only will the child be free of the disease in question, but also the child's genome will be repaired. In other words, germ-cell gene therapy has the potential to eradicate the mutation from the parents' genome, the result of which is that neither the child nor the child's progeny will be afflicted with the disease. If all people with the genome mutation receive germ-cell gene therapy, the mutation will no longer appear in the human genome and the disease will be eradicated. The exception to this is in the case of a disease that results from spontaneous genome mutation. Thus, the potential for disease eradication and the benefits such eradication would have for society as a whole justify gene therapy research in this Utilitarian argument.

The argument for reproductive freedom is best viewed from the patient's perspective. In a case where both parents are carriers for a disease, germ-cell gene therapy may be their only opportunity to have healthy children. It is possible to envision a couple who would be willing to undergo germ-cell gene therapy, yet unwilling to terminate a pregnancy if genetic mutation is detected during gestation. In such a circumstance, withholding potential preventive technology appears unethical. The intervention of germ-cell gene therapy is performed on two consenting adults, not a fetus, an entity from whom informed consent cannot be obtained. Those concerned with the rights of a fetus or a child should support germ-cell gene therapy because it mitigates the need for invasive procedures for the child (such as somatic-cell gene therapy) and abortion. In fact, germ-cell gene therapy is likely to significantly minimize if not completely eliminate the need for aborting genetically impaired fetuses. An extension of reproductive rights for the parents is the right of the child to be born with optimal health. If it is possible for the medical community to ensure that a child is born without genetic defect, then the burden of providing this treatment is an inherent responsibility.

These arguments merely graze the surface of the kaleidoscope of supporting arguments available for gene therapy. They do represent gene therapy advocacy from both the communal and individual perspective. In summary, gene therapy has both palpable and indirect potential benefits for society as a whole (*via* the treatment of genome mutations and possible disease eradication) and individuals (both parents and potential offspring).

3. ARGUMENTS AGAINST GENE THERAPY RESEARCH

Gene therapy has a strong opposition, based on well-founded arguments and concerns. Four strong criticisms of gene therapy research are its irreversible nature, the potential effects on the gene pool, informed consent (or lack thereof), and risk of harm. The first two critiques view the potential ramifications of gene therapy research from a societal perspective, while the second two focus more on the individual(s) for whom the technology will be used. This is by no means a comprehensive list of the dissenting arguments, they are merely demonstrative of the philosophy which does not support gene therapy research.

One of the most common concerns about gene therapy is its irreversibility. Genetic mapping is a relatively new advancement in science and therefore is not fully understood. This leads to concerns about eliminating any occurrence in the human genome. For example, if mutation X in an individual's genome was known to cause a specific disease, germ-cell gene therapy could theoretically be used to 'repair' the mutated section of that individual's genome. In so doing, the individual's genome would be permanently altered to what current scientific knowledge considers good or correct. The decision to perform this alteration affects the individual treated and all of his/her offspring indefinitely. The concern with this procedure is the potential that the portion of the genome that is currently considered pathological may actually have unknown positive effects on the individual. Consider the following hypothetical: think back to the time when sickle-cell anemia was identified and assume that the genetic mutation that causes sickle-cell anemia was identified at the same time. Now, consider what would have happened if the germ-cell gene therapy had been a generally accepted medical technique at the time. Germ-cell gene therapy would have been used to 'repair' the portion of inflicted individual's genome that coded for sickle-cell anemia. It would have likely been considered a great success in medicine to completely eliminate the sickle-cell anemia mutation from the human genome. The problem with the scenario is that we have since learned that the mutation for sickle-cell anemia has great benefit. Specifically, individuals who carry the sickle-cell anemia mutation are less susceptible to contracting malaria. Had the sickle-cell anemia mutation been identified and eradicated it is likely that scientists would never have discovered this resistance to malaria. It is also possible that tremendous portions of the African population who are carriers of sickle-cell anemia would have been wiped out by malaria. Furthermore, scientific advancements that have been inspired or structured from the sickle-cell anemia/malaria link would never have been made. Thus, it is legitimate to be concerned about utilizing germ-cell gene therapy to eliminate a portion of human genome that scientists currently believe to be bad because further scientific discovery may reveal that, in fact, the mutation had positive benefit for its carrier.

Another significant concern expressed by the opposition to gene therapy research is the concept that germ-cell gene therapy will ultimately diminish the human gene pool by removing genes considered to be bad. This is similar to the previous argument about irreversibility, in which the concern was that the germ-cell gene therapy would remove portions of the human genome that may ultimately turn out to be beneficial. The diminished gene pool concept looks more broadly at what is lost when the gene pool is reduced. Many, particularly in Europe, believe that there is a right of all unborn children to be born with their entire genome intact. To view it another way, scientists have an obligation to preserve the diversity of the gene pool [5]. To diminish the gene pool manually, i.e. through scientific advancement, is un-Darwinian or unnatural. Those who oppose germ-cell gene therapy on these grounds must also oppose pre-implantation genetic diagnosis of fetuses used for in vitro fertilization and prenatal genetic diagnosis for the purpose of selective abortion because these techniques also serve to diminish the gene pool. Both the irreversibility and diminished gene pool arguments are made from a Utilitarian perspective on the assumption that eradicating elements of the human genome may in the future be detrimental to a substantial portion of the population.

When the opposition approaches gene therapy from an individual's perspective, the greatest concern is a lack of informed consent. This argument is not applicable to somatic-cell gene therapy because the treatment is performed on each individual (who can provide informed consent prior to treatment) and affects only that individual. Contrarily, the germ-cell gene therapy may not directly affect the individual who undergoes the treatment, rather it is designed to benefit the individual's offspring and all future generations. Germ-cell gene therapy would be performed on the parent prior to conception of the offspring, thus making it impossible for the offspring to consent. It is significant to note that the effects of germ-cell therapy reach beyond the treated individual's children, in fact an altered genome will be inherited by all future generations that descend from the treated individual.

Arguably the strongest argument in opposition to gene therapy research in humans is the tremendously high risk for complications. The risk that separates gene therapy from all other innovative therapies is the method by which genetic material is transferred. In somatic-cell gene therapy, active vectors (most common are adenoviruses and retroviruses) are used to transfer genetic material. The vectors have the capacity to potentially propagate on their own [6] and may become active in areas other than where they were inserted [7]. These agents may have long latencies, delaying detection of the mutation and the ensuing immune response. Another element of risk is derived from gene therapy trials, which were performed using transgenic animals that produced high rates of mutagenicity. The lack of success in animal models implies that there is a high level of risk in human trials. Proponents of this argument believe that, at the very least, human subject trials should not be performed until after the techniques have been consistently successful in animal trials. In addition to the mutagenicity observed in animal models, the concern is compounded by the frequency with which complications are occurring in current human subject trials [8]. Examples of gene therapy trials that have encountered negative outcomes are the death of Jesse Gelsinger at the University of Pennsylvania [9] and the study subjects who developed leukemia secondary to insertional mutagenesis in the French gene therapy trials [10].

As with pro-gene therapy arguments, these four arguments against gene therapy are only the beginning of a plethora of organized opposition to gene therapy research. As demonstrated here, this point of view can be substantiated with both individual and societal concerns. In summary, gene therapy poses very real risks to the individual being treated, their progeny and society as a whole.

4. GENERAL SOCIETAL CONCERNS REGARDING GENE THERAPY RESEARCH

Beyond the heated debate over gene therapy research in the medical community is the same debate in society at large. Public examination of the issues involved in gene therapy research can be found in venues of politics and *academia*, but discussions on the topic can also be found in newspapers, public radio, and television. Dialogs reviewing the ethical merits of this research, typically in the form of benefit versus risk, have become commonplace. Some of the common concerns expressed include issues of distributive justice, risk of eugenics, the concern of playing God, and the slippery slope argument.

Many have expressed concern that the principle of distributive justice will not be honored in the case of gene therapy [11,12]. This is the fear that only rich populations will benefit from gene therapy due to the prohibitive cost of the treatment. The disparity between treatment and no treatment has the potential to foster discrimination from two perspectives. If an individual with a genetic disability is unable to afford gene therapy, he may suffer discrimination at the hands of a society that has 'cured' others of the same affliction. A second scenario is one in which an individual with a genetic disability chooses not to undergo gene therapy, similar discrimination may occur. The counter argument to the theory that gene therapy should not be permitted because it will surely violate the principle of distributive justice is the argument that many if not all of medical technology violates the principle of distributive justice, yet society does not prevent other medical advances. In other words, justice is not served by forcing one person to suffer because another person must suffer.

For many, manipulating the human genome *via* gene therapy rekindles fears of eugenics. Eugenics, the science of attempting to improve the human genome by eliminating 'bad' genes, is reminiscent of Nazi efforts to create a superior human race and US-mandatory sterilization laws. In this case, the fear of gene therapy stems from a vision of the future in which people can 'engineer' their fetus to produce their ideal child. With the advancement of gene therapy, it is foreseeable that science could allow parents to choose their child's hair or eye color, IQ, or physical abilities. If this advanced ability for genetic manipulation were attained, ethical landmines would abound in issues of distributive justice (only the rich could afford to design their ideal child) as well as discrimination and narrowing of the gene pool. The counterarguments to the potential eugenic uses of gene therapy are twofold. First, current gene therapy research is searching for cures or preventions to disease, not seeking the capacity for pure enhancement. Second, under this broad definition, eugenics is currently being practiced in the form of pre-implantation genetic diagnosis for *in vitro* fertilization, prenatal diagnosis in combination with selective termination, and genetic testing prior to marriage.

Another commonly heard argument against human gene therapy is the belief that it is unethical because tampering with the human genome is transgressing into God's territory. In other words, many believe that altering an individual's genome is 'playing God'. This is by far the most difficult perspective to prove and/or disprove because it is not a deducible fact, rather a religious belief. The belief that the human genome is solely for God's purview arises from a religious theory called genetic determinism, a theory in which each individual's genetic composition is predetermined by God. Substantiation for the argument that gene therapy is 'playing God' is based in religious philosophy and therefore cannot be assembled into a secular argument. Those outside the religions that subscribe to the theory of 'playing God' question how gene therapy is more intrusive into God's purview than a heart transplant, sex change operation, or separation of Siamese twins.

Possibly the most common concern voiced about gene therapy research is the slippery slope argument. The theory of the slippery slope argument is that once you begin down a certain path of knowledge, even if the purpose of the path is righteous and ethical, the information gained during travels on that path can and will ultimately be used for a harmful purpose. In other words, once knowledge is gained it cannot be unlearned and therefore might and likely will be used for a questionable purpose. The specific fear in the slippery slope of gene therapy is that once scientists learn how to manipulate the human genome to prevent and cure disease, the same knowledge can and will eventually be used unethically for purposes such as enhancement. The counterargument to the slippery slope theory is best stated by Gorovitz, "fortunately, it is possible to start down a slippery slope and then to stop."[13]. This is clearly the case given the state of many current medical therapies that are at the top of a slippery slope. For example, a woman can choose to have elective surgery that renders her incapable of having children (i.e. a tubal ligation). This procedure is performed on a regular basis throughout the country, yet the original fear that this procedure would ultimately be used to sterilize women against their will (a form of eugenics) has rarely materialized.

Thus, society has many different methods of examining the question of gene therapy research. One can approach the gene therapy debate from religious, political, and personal avenues. It is significant to recognize that none of the arguments, either in favor of or against gene therapy research, are significantly different from any other area of scientific advancement. In other words, human gene therapy is not unique, rather it is subjected to the same criticism and support as all other medical therapies during their initial introduction into general use.

5. THE AMERICAN APPROACH TO GENE THERAPY REGULATION

The United States pioneered the field of human gene therapy as well as the ethical safeguards surrounding its research. The United States developed an independent organization called the Recombinant DNA Advisory Committee (RDAC), whose primary purpose is to review human gene transfer study protocols in the United States. In addition, the RDAC "organizes safety conferences and publicly maintains databases and adverse event reports, thus enabling investigators to design safer trials." [14]. The RDAC is a central review body, as opposed to research ethics boards which are institution-specific. The universal nature of a central review board is in the best position to look at human gene transfer research from a panoramic perspective [15]. From its universal vantage point, the RDAC is uniquely situated to air, debate and resolve conflicting ethical, social and safety claims [16].

There are seven generally accepted ethical requirements for a research trial: (1) value, (2) scientific validity, (3) fair subject selection, (4) favorable risk–benefit ratio, (5) independent review, (6) informed consent, and (7) respect for enrolled subjects [17]. Emanuel goes on to say that the two greatest challenges for an ethics committee are to ensure that the researchers performed adequate risk disclosure during the consent process and that the proportionality of risk and possible benefit were adequately weighed. The conclusion of the US's universal ethics review board is best summarized by Theodore Friedmann who states that all of these questions and potential problems in human gene therapy research should not thwart or stagnate attempts to further scientific knowledge. He states, "To make progress, one must accept the limitations of knowledge and simultaneously use available information to ease suffering and to continue research into improvements in technology." [18].

REFERENCES

[1] B. K. Zimmerman, Human germ-line therapy: The case for its development and use, *J. Med. Philos.,* 1991, **16**, 593–612.

[2] J. C. Fletcher and W. F. Anderson, Germ-line therapy: A new stage of debate, *Law, Med. Health Care*, 1992, **20**(Spring–Summer), 26–39.

[3] E. M. Berger and B. M. Gert, Genetic disorders and the ethical status of germ-line gene therapy, *J. Med. Philos.*, 1991, **16**, 667–683.

[4] I. Szebik and K. C. Glass, Ethical issues of human germ-cell therapy: A preparation for public discussion, *Acad. Med.*, 2001, **76**, 32–38.

[5] R. Munson and H. D. Lawrence, Germ-line gene therapy and the medical imperative, *Kennedy Inst. Ethics J.*, 1991, **2**, 137–158.

[6] D. Williams, Clarity and risk: The challenges of the new technologies, *Med. Device Technol.*, 2001, **12**, 12–14.

[7] J. Kimmelman, Recent developments in gene transfer: Risk and ethics, *BMJ*, 2005, **330**, 79–82.

[8] J. Savulescu, Harm, ethics committees and the gene therapy death, *J. Med. Ethics*, 2001, **27**, 148–150.

[9] S. Jacein-Bey-Abina, LMO2-associated clonal T cell proliferation in two patients after gene therapy for SCID-X1, *Science*, 2003, **302**, 415–419.

[10] T. Tannsjo, Should we change the human genome? *Theoret. Med.*, 1993, **14**, 231–247.

[11] J. Mendeloff, Politics and bioethical commissions: "Muddling through" and the "Slippery slope", *J. Health Politics, Policy Law*, 1985, **10**(Spring), 81–92.

[12] National Institutes of Health, Appendix L, in *NIH Guidelines for Research Involving Recombinant DNA Molecules*, Department of Health and Human Services, Washington, 2001.

[13] J. Kimmelman, Protection at the cutting edge: The case for central review of human gene transfer research, *CMAJ*, 2003, **169**(8), 781–782.

[14] A. Gutmann and D. Thompson, Deliberating about bioethics, *Hastings Cent. Rep.*, 1997, **27**(3), 38–41.

[15] E. J. Emanuel, D. Wendler and C. Grady, What makes clinical research ethical? *JAMA*, 2000, **283**(20), 2701–2711.

[16] T. Friedmann, Principles for human gene therapy studies, *Science,* 2000, **287**(5461), 2163–2165.

Do Germline Interventions Justify the Restriction of Reproductive Autonomy?

Johannes Huber

Faculty of Medicine, Ludwig-Maximilians-University, Munich and Institute for Scientific Issues Related to Philosophy and Theology, School of Philosophy S.J., Munich, Germany

Abstract

Gene therapy in human beings might carry severe risks for subsequent generations. The health of future offspring can be endangered by not only intended interventions in germline cells, but also by inadvertent effects of somatic gene therapy. Therefore, one should investigate whether interventions into the germline might justify a restriction of reproductive autonomy in order to limit possible side effects to the treated individual. In a first step, relevant germline interventions and their risk profile will be presented briefly. Subsequently, a definition of "reproductive autonomy" and the development of different methods for a gradual limitation of this fundamental right will enable us to discuss the legitimacy of such a restriction. By pointing out general ethical aspects and probable further developments, some considerations concerning the justification of restricting reproductive autonomy can be made. Some measures adopted so far will be illustrated, and a short summary of the actual discourse about the relevance of inadvertent germline effects will show the difficulties regarding risk assessment in this special case. It will then be suggested that, at this point, a risk-adjusted approach for restricting individual reproductive autonomy seems to be the best possible solution.

Keywords: Gene therapy, germline, reproductive autonomy, risk assessment, social coercion.

Contents

1. Introduction 110
2. What kind of germline interventions should be investigated and what are some of the related risks? 111
   2.1. DNA level 112
   2.2. Level of epigenetic regulation 112
   2.3. Cellular level 113
3. A definition of "reproductive autonomy" and possible degrees of its limitation 113
   3.1. Influence aimed at public opinion 114
   3.2. Active counselling of individuals in question 114
   3.3. Regulations by law 115
   3.4. Sterilization as a condition for treatment following informed consent 115
4. Discussion 116
   4.1. Somatic cell gene therapy after birth 117
   4.2. Gene therapy before the differentiation in soma and germline cells 118
5. Summary 119
References 120

1. INTRODUCTION

Transmission of hereditary diseases to the next generation by coital reproduction cannot be subject to legislation in democratic societies because procreative liberty is considered a basic human right. Privacy and freedom of reproductive issues have to be respected, and no autonomous individual can be forced to abstain from reproducing. The importance of this fundamental right becomes particularly obvious when it is absent. In totalitarian regimes, reproduction typically is a major political concern: people are told whether or not to reproduce, with whom to reproduce and how many children to have. This policy can be pursued by means of propaganda, social coercion or brutal force. In overcoming the nationalization of one of the most private issues in life, reproductive autonomy has become a precious good that has to be protected.

Ideally, human reproduction is only subject to the moral judgement of those involved, not to social coercion. Coherently, parenthood naturally conceived cannot be considered to be an injury towards offspring. The main feature of this assessment is the nativeness of human reproduction because the course of nature is not to blame. But as soon as techniques such as prenatal diagnostics combined with the possibility of abortion or complex interventions such as assisted reproduction come into play, additional actions are involved in the process of human reproduction making it artificial to a certain extent [1]. Although a child conceived in the course of such a medical treatment must never be considered to be "produced", one has to keep in mind that human actions were involved in a formerly all-natural process and that those manipulations must comply with a defined standard.

Thus, a new quality of responsibility arises, and lately even terms such as warranty or guilt have been used in the context of human reproduction. For example, courts are confronted with claims about "wrongful births": parents demand compensation for the birth of a disabled child by claiming that the pregnancy would not have been continued but for the negligence of doctors who failed to fully inform them [2]. Obviously, a great sense of responsibility is mandatory when additional human actions are taken in reproduction in order to prevent possible harm to the subsequent generations.

A similar concern might be given to another innovative technique in the near future, when gene therapy in humans becomes more frequent, and possible risks for the germline cannot be excluded. Not only intended interventions in germline cells might endanger otherwise healthy offspring, but also possible side effects of somatic gene therapy could cause severe diseases in future offspring. For this reason, one should investigate whether interventions into the germline might justify a restriction of reproductive autonomy in order to limit possible side effects to the treated individual. In a first step, relevant germline interventions and their risk profile will be presented briefly. Subsequently, a definition of "reproductive autonomy" and the development of different methods for a gradual limitation of this fundamental right will enable us to discuss the legitimacy of such a restriction. In this context, the present German legislation concerning germline interventions will be examined. This is aimed to show that a precise risk assessment is necessary to judge whether an appeal to the personal responsibility of parents is sufficient or not. Is a governmental measure required to protect the well-being of the next generation from the side effects of human gene therapy?

2. WHAT KIND OF GERMLINE INTERVENTIONS SHOULD BE INVESTIGATED AND WHAT ARE SOME OF THE RELATED RISKS?

Although there is a disagreement about the assessment of human gene therapy in general, two major ethical landmarks are widely accepted. The first one concerns the aim of the intervention, distinguishing prevention or cure of a defined disease from enhancement [3]. The second essential point for ethical evaluation is the range of the treatment: Are the effects of a genetic manipulation limited to a single person or might there be an impact on a theoretically unlimited number of persons [4]? The latter is the pivotal criterion discussed here.

As we investigate how gene therapy-associated risks could be limited to the patient being treated, the usual meaning of "germline therapy" will not be taken into account because its objective is to affect offspring. Therefore, the procedures in question are (i) somatic cell gene therapy and (ii) gene therapy in early developmental stages when the differentiation in soma and germline cells has not yet taken place.

Both interventions are aimed at curing the treated person only, and their basic idea is identical [5]: A mutated gene causing a certain disease ought to be substituted by inserting an intact version of this gene. For this purpose, biological vehicles or physical methods are needed to transport the therapeutic gene inside the nucleus of the cells, which express the pathological gene. Nowadays, genetically modified viruses are the preferred system for gene delivery in clinical trials as they are able to invade cells actively, thereby introducing their own genetic material. This method exhibits the highest *in vivo* transfection efficiency. Albeit, only somatic cells are envisaged, the vector viruses are not yet acting specifically and might therefore also affect the germline [6,7]. To find an alternative and more specifically acting vectors is a matter of great importance [8], but difficulties in achieving sustained gene expression by safe and efficient gene delivery still exist [9].

Another important fact when assessing the risk of germline modifications in somatic gene therapy is the developmental stage at the time of intervention. While somatic gene therapies proposed and already conducted in adults and children seem to bear a relatively low risk of altering germline cells, interventions conducted earlier in individual development seem more likely to render this result. Since prenatal and even preimplantation diagnosis can be performed, therapy in those early stages of life seems to be a logical subsequent step. *In utero* therapies may involve a higher risk of germline modifications, which cannot sufficiently be quantified yet [10]. Moreover, going back even further in the ontogenetic development, a distinction between somatic and germline gene therapy is not possible anymore as this differentiation is not established until the primordial germ cells are formed during the second week of development [11]. Therefore, every efficient gene therapy performed during the first two weeks of individual development will equally affect both germline and soma cells.

Unlike common side effects of gene-therapeutic treatments such as allergic or inflammatory reactions [12] that are caused by interactions with the used vector, the whole spectrum of genetically caused problems can be transmitted to subsequent generations if the germline is affected [13]. Most of these risks can be categorized according to the fact that a defined DNA sequence works only in a given context established on three main levels.

2.1. DNA level

Even when the DNA sequence in question is successfully transferred into the cell nucleus and no mutation takes place during this process, one major problem remains at the DNA level. It is crucial that the therapeutic gene is implanted in the correct place in order for the gene to function faultlessly. Otherwise, a phenomenon known as insertional mutagenesis is likely to happen, in which the ectopic chromosomal integration of a DNA sequence either disrupts the expression of a tumour-suppressor gene or activates an oncogene leading to the malignant transformation of cells [14]. In other words, integrating a gene in the wrong place can cause the development of cancer. Consequently, as long as genes are not securely and accurately inserted into a defined place, the treated patient and – if the germline is affected – also the children are at high risk of malignant transformations.

2.2. Level of epigenetic regulation

Regulation of gene expression is an extremely complex process that is currently being researched and defined. A lot of different mechanisms implement the expression of a cell type-specific subset of genes at a certain time and in a defined quantity. As this interpretation of the genome happens without altering it, epigenetic mechanisms can be defined as changes in gene function that are not based on changes in DNA sequence. Covalent chemical modifications of the DNA (DNA methylation) and of its packaging (histone and chromatin modifications) are the most important mechanisms known so far.

This level of regulation seems to be sensitive even to interventions on a cellular level such as intracytoplasmatic sperm injection (ICSI) – a technique in assisted reproduction where a spermatozoon is actively transferred into the oocyte by microinjection. As such manipulations seem to result in an increased number of diseases caused by epigenetic defects [15,16], gene therapy could entail similar consequences. Introducing such damages into the germline would lead to an accumulation of severe epigenetic diseases that are relatively rare today [17].

2.3. Cellular level

Finally, oncogenesis – although sometimes caused by definable mutations on the DNA level – is a cellular phenomenon that arises from complex interactions among the genome, the organism and the environment. Multifactorial coherences like this might easily be influenced by side effects of gene therapy. Moreover, inserting artificial chromosomes and plasmids could interfere with essential cellular processes such as mitosis and meiosis, which would be lethal in the early stages of development. Other possible risks might be early ageing and diseases due to altered chromosome architecture.

In summary, germline interventions involve risks for patient offspring that must not be disregarded. Thus, in order to assess a possible restriction of reproductive autonomy with the view of limiting these risks, we first need to clarify the meaning of the term.

3. A DEFINITION OF "REPRODUCTIVE AUTONOMY" AND POSSIBLE DEGREES OF ITS LIMITATION

Discussions of the meaning and plausibility of reproductive autonomy have gained growing importance in connection with artificial reproductive techniques, prenatal diagnostics and abortion.

Whether one is willing to reproduce or not, must be regarded as a fundamental privilege or claim of decision for everybody. This implies that there is no moral duty that can be violated by either decision, and that others – particularly the state – are morally bound not to interfere with the exercise of this liberty unless there are compelling reasons for restricting it. This freedom exists as long as the decisions taken do not result in substantial harm for others [18,19].

Reproductive autonomy can be understood to be a particular form of autonomy, a capacity or power inherent in every single person. Acting autonomously can stand for (a) having the power of self-government and (b) exercising this power [20]. Employing the notions of ability and opportunity can further develop this concept of autonomy: in order to have the power of self-government, someone needs both the ability and the opportunity

	Power of self-government	Exercise of this power
Ability	Internal constraints making us unable to decide	Internal constraints making us unable to execute our decision
Opportunity	External constraints making us unable to decide	External constraints making us unable to execute our decision

Fig. 1. Possible constraints on reproductive autonomy.

to decide about performing a certain action. Similarly, in order to exercise this power, someone needs both the ability and the opportunity to realize the decision made [21]. Someone's ability can be reduced by internal constraints, whereas the opportunity to do something might be controlled externally. To restrict reproductive autonomy can, therefore, be governed by external as well as internal constraints. The following scheme of reproductive autonomy (see Fig. 1) might facilitate classification of the possible constraints given below.

Every measure adopted for limiting individual reproductive autonomy must tend to protect subsequent generations from harm by preventing defective gametes from being transmitted and attempt to avoid early embryos being affected during pregnancy. Thus, before any gene-therapeutic intervention, gametes should be saved and stored, and contraception should be guaranteed as far as possible after the intervention. In a constitutional state, four approaches for limiting reproductive autonomy seem feasible; in the following scheme both the degree of coercion [22] and the security grow, starting from the most moderate measure. All alternatives can be combined amplifying each other.

3.1. Influence aimed at public opinion

As soon as gene therapy in human beings becomes a more common therapeutic alternative, influencing public opinion will surely be necessary. Objective information without other underlying interests would be the ideal form of transporting knowledge about general risks and about how to avoid handicapped children after gene-therapeutic interventions. If this approach is not abused, it will extend the ability to come to a well-founded decision rather than to restrict reproductive autonomy. The weak point of influencing the public opinion is that normally the impact on individual behaviour is quite limited.

3.2. Active counselling of individuals in question

Active counselling is a widely used practice in medicine, especially in human genetics [23]. A good example in our context is the recommendation given to patients, who want to undergo chemotherapeutic treatment, to desist from conception for a certain period of time, as germ-cell affections cannot be excluded, and to use appropriate contraceptives.

Expert counselling offered directly to affected patients is surely of greater influence than general information offered to the public, as it can be more personal and adjusted to the individual situation. Depending on how directive the counselling is, its impact on the parents can range widely: from enabling to decide well informed but still autonomously, to an external and internal constraint on the power of self-government.

3.3. Regulations by law

Counselling can also be dictated by law, e.g. as it is conducted in the German regulation of abortion. A far-reaching coercive measure is to prohibit reproduction under defined circumstances. An existing and widely accepted restriction of reproductive autonomy by law is the incest taboo: in order to protect the following generations from harm caused by an aggravation of the individual genome, a social and legal rule limits free choice of a partner. Whether a similar legislation against vertical transmission of germline defects could make sense and can be justified has to be discussed. This degree of state control would mean a strong constraint on all the qualities of reproductive autonomy.

3.4. Sterilization as a condition for treatment following informed consent

Patients in need of a certain gene-therapeutic intervention that involves a very high risk for an affection of the germline could be asked to undergo sterilization as a condition for treatment. As they hope to profit from the planned intervention, they have to take responsibility not only for themselves but also for their subsequent generations. Therefore, sterilization might be suitable especially for cases that seem too dangerous to be performed otherwise.

Although forced sterilization under any circumstance is inconsistent with the liberal foundation of a democratic constitutional state, a voluntary decision for such a treatment involved in an informed consent seems thinkable. As this design does not rely on a direct governmental action, it is less coercive than a regulation by law. But given the finality of the proposed measure it must be considered an extreme option. Under ideal conditions, an informed consent would restrict neither the power of self-government nor the exercise of this power as the decision to undergo the treatment is voluntary, and it can be performed freely.

Each of these four options has already been engaged in regulating possible risks of gene therapy. By discussing general ethical aspects and probable further developments, full consideration must be given about how restricting reproductive autonomy can be justified. In the following sections, some measures adopted so far will be illustrated, and a short summary of the actual discourse about the relevance of inadvertent germline effects will show the difficulties of risk assessment in this special case.

4. DISCUSSION

Some basic points are important for discussing adequate measures for limiting the risk of inadvertent germline effects.

In view of all the risks possibly associated with human gene therapy, some authors argue for a general prohibition of such medical treatment. According to the "imperative of responsibility" [24], a new method, in which severe side effects cannot be excluded securely in the long run, must not be employed because of the grown responsibility towards future generations due to extremely powerful modern technologies. This ethical principle seems to offer the highest degree of security; however, the duty to help suffering people is neglected. Performing a possibly dangerous therapeutic treatment can have negative effects, but are there not also negative effects to omitting research and impeding the development of novel techniques? Where careful risk assessment shows minimal danger, it would be ethically indefensible to prohibit somatic gene therapy when patients' lives could be saved or significantly enhanced. Therefore, a careful consideration of chances and risks seems more appropriate.

For example, the question has to be posed whether there is a difference in meaning between naturally occurring mutations and artificially caused changes in human germline cells. As already considered in the very beginning of this chapter, human actions influencing natural processes like reproduction are of great ethical relevance. Without committing a naturalistic fallacy, natural changes in human genomes are less problematic than artificial ones because, unlike spontaneous natural events, the latter represents an action somebody can be held responsible for. This responsibility exists for intended and accidental changes likewise, and is not attenuated in principle by actual findings about the plasticity of the human genome [25]. Although our former understanding of a quite stable transmission of genetic material might be wrong to a certain extent, the frequency of endogenous mutations in humans is not a valid argument for lowering our standards in protecting the germline. As one natural endogenous genomic insertion is estimated in every 50–100 individuals [26], it was argued that a limit of less than one event in 6000 germ cells set by the US American Federal Drug Administration (FDA) should be reconsidered [27]. This line of argumentation neglects the different qualities of naturally occurring and artificially caused mutations. Therefore, a cardinal point in this discussion must not be to measure the dimension of insertion tolerable in human gene therapy trials by the frequency of naturally occurring events, but to minimize the risk for offspring through regulation [28].

This maxim is especially important, as artificially induced diseases once established cannot be controlled sufficiently. The problem of control over the spreading of a new founder mutation – a mutation carried by an individual or a small number of people who are among the founders of a future population – might be feasible for technical reasons; however, not for ethical ones. Although a certain class of "mutated" human beings might successfully be restrained from reproducing, such a proceeding can never ever be compatible with fundamental human rights. As every human being is born equally and therefore entitled with the same rights, a separation of a genetically modified subgroup contradicts human dignity.

An important aspect favouring safety and health of the next generation is the future parents' benevolence towards their children: if they know about a severe risk for their offspring, most of them will take a reasonable decision. This underlying assumption is fundamental for successful counselling and may be taken into account when reasoning about appropriate measures for limiting the transmission of inadvertent germline effects. Another problem is the restriction of intended germline modifications, as those could be understood as enhancement, thereby encouraging parents to pass on certain features to the next generation.

Keeping these considerations in mind, two forms of treatment can be investigated and adequate regulations can be proposed. The first one represents an actually given situation, while the latter develops a future scenario.

4.1. Somatic cell gene therapy after birth

In the United States, an exemplary regulation has already been established by the FDA enabling a description of how reproductive autonomy can gradually be restricted according to the estimated risk[1] [29,30].

Risk assessment is a scientific question and must be established on a case-by-case basis requiring close investigation of therapy protocols. Therefore, if a new vector is used, or if reliable data about biodistribution and other safety concerns are not available for other reasons, patients are required to be sterile in order to take part in clinical trials. This rule led to the situation that some patients suffering from haemophilia volunteered to be sterilized before gene transfer experiments, just to be allowed to take part in the trial [27]. In sum, this regulation matches level 4 of the above given methods for limiting reproductive autonomy. Such a strict rule might be justified by the existence of an incalculable risk threatening future generations.

---

[1] Proposed *in utero* treatments are not discussed here as another contribution concentrates on fetal gene therapy.

At later stages, when basic investigations have been conducted in order to identify vector receptors and to assess the likelihood of germ-cell transduction, all features of a given vector necessary for assessing its risks of inadvertent germline effects should be known. If this process leads to the scientific judgement that the risk of affecting the germline is extremely low [30,31], therapies involving the characterized vector may be performed on a less restrictive level. For this type of "established" somatic gene therapy, level 2 seems appropriate. Participants in such a trial should be informed of the risk of germline alteration and ways to minimize this risk: "To avoid the possibility that any of the reagents employed in the gene transfer research could cause harm to a foetus/child, subjects should be given information concerning possible risks and the need for contraception by males and females during the active phase of the study. The period of time for the use of contraception should be specified. The inclusion of pregnant or lactating women should be addressed" [32].

This kind of risk-adjusted restriction of reproductive autonomy seems to be the best solution possible at the moment. It manages to minimize potential risks, pays respect to reproductive freedom and maintains the chances of further development at the same time.

4.2. Gene therapy before the differentiation in soma and germline cells

An interesting future scenario is the possibility of gene-therapeutic treatment for preimplantation embryos. Although the scientific development of preimplantation genetic modification (PGM) in humans might be ethically questionable [33], a different situation might arise after the establishment of such a technique. This might provoke an ethical dilemma, for example, in Germany where legislation considers even early embryos to be endowed with the right to life and human dignity.

If in an early embryo a severe genetical defect was found that could be treated by PGM and that otherwise would inevitably lead to early death, a strong ethical obligation could be felt to cure this human being. On the other hand, gene therapy at this early stage of development would almost surely lead to an alteration of germline cells and would therefore endanger the offspring as long as the secure transfection does not exist. Given the actual legislation which prohibits intended germline effects [34], a treatment in favour of the developing child would be a punishable act only because the differentiation in germline and soma cells has not happened yet.

Apart from the fact that the whole scenario is highly theoretical, it would be impossible under current German law as preimplantation genetic diagnosis (PGD) is prohibited due to the "Act for Protection of Embryos" (EPA) [35]. But our example reveals a weakness of the latter. The EPA prohibits intended germline alterations, but explicitly excludes inadvertent germline effects: "Anyone who artificially alters the genetic information of a human germ line cell will be punished with imprisonment up to five years or a fine" [35]. But this statement does not apply to "inoculation, radiation, chemotherapeutic or other treatment by which an alteration of the genetic information of germ line cells is not intended" [35].

A regulation that ensures the sterility of the patient after the treatment (level 4) could solve the problem. As an informed consent is not possible in the given situation, regulations must be decided according to the presumed will of the patient. The alternatives are quite clear: death or the chance to live a healthy life without genetically related children. A possible approach could be the synchronous transfection of the early embryo with the curative DNA sequence and another gene causing infertility. Thereby, germline modifications would not be the intended aim and could not cause any harm.

In the light of this scenario, the German EPA seems to be too strict on the one hand and careless on the other. How much can good intention not to alter the germline help those human beings suffering from diseases caused by mere side effects? What about our responsibility towards future generations? And why should a therapeutic intervention in germline cells be prohibited although transmission to subsequent generations can be excluded? Therefore, not only the intention of an action alone should be subject to legislation, but also the consequences of performing a treatment.

In both scenarios, some common future achievements can be claimed: further research, especially in suitable animal models, is crucial for investigating possible risks and their quality and quantity. Another important means for exploring side effects of treatments performed is a close and detailed long-term follow-up [29] of patients and their offspring. Moreover, interdisciplinary [36,37] and prospective [38] approaches on evaluating the results of scientific research should be encouraged.

5. SUMMARY

While pursuing deliberate germline modifications in humans in order to treat offspring would be inappropriate given the current technology; the risk of inadvertent germline modifications must be taken seriously. Given the possible benefits of such therapies, this problem has to be solved.

The risk of inadvertent germline effects in actual trials of somatic cell gene therapy is thought to be very low, but is still hard to be assessed exactly. Therefore, a risk-adjusted approach for restricting individual reproductive autonomy seems to be justified.

In summation, maybe the problem ends in the inverse question: Can the restriction of reproductive autonomy in certain cases enable germline interventions to be a therapeutic alternative with calculable risk – at least as long as negative consequences for forthcoming generations are likely to happen?

Johannes Huber (born 1980) graduated from the Ludwig-Maximilians-University Medical School and from the School of Philosophy S.J., both in Munich. He just finished a dissertation in medicine about theoretical aspects of the oocyte-to-embryo transition and is preparing his doctoral thesis in philosophy at the "Institute for Scientific Issues Related to Philosophy and Theology" focusing on the meaning of "totipotency". His principle research interests include philosophy of nature, bioethics and ethical issues in medicine.

REFERENCES

[1] A. Kuhlmann, Reproduktive Autonomie? Zur Denaturierung der menschlichen Fortpflanzung, *Deutsche Zeitschrift für Philosophie*, 1998, **6**, 917–933.
[2] N. Priaulx, Joy to the world! A (healthy) child is born! Reconceptualizing 'harm' in wrongful conception, *Soc. Legal Stud.*, 2004, **13**, 5–24.
[3] W. F. Anderson, Human gene therapy: Why draw a line? *J. Med. Phil.*, 1989, **14**, 681–693.
[4] E. -L. Winnacker, T. Rendtorff, H. Hepp, P. H. Hofschneider and W. Korff, Gene technology: Interventions in humans. *An Escalation Model for the Ethical Evaluation*, 4th edn, Herbert Utz, Munich, 2002.
[5] L. Walters and J. G. Palmer, *The Ethics of Human Gene Therapy*, Oxford University Press, Oxford, NY, 1997.
[6] V. R. Arruda, L. Couto, D. Leonard, K. Addya, J. Sommer, R. W. Herzog, M. A. Kay, B. Glader, C. Manno, A. Chew and K. A. High, Risk of inadvertent germline transmission of vector DNA following intravascular delivery of recombinant AAV vector, *Mol. Ther.*, 2002, **5**, s159–s160.
[7] A. S. Pachori, L. G. Melo, L. Zhang, M. Loda, R. E. Pratt and V. J. Dzaua, Potential for germ line transmission after intramyocardial gene delivery by adeno-associated virus, *Biochem. Biophys. Res. Commun.*, 2004, **313**, 528–533.
[8] D. J. Glover, H. J. Lipps and D. A. Jans, Towards safe, non-viral therapeutic gene expression in humans, *Nat. Rev. Genet.*, 2005, **6**, 299–310.
[9] A. Rolland, Gene medicines: The end of the beginning? *Adv. Drug Deliv. Rev.*, 2005, **57**, 669–673.
[10] L. A. Caplan and M. J. Wilson, The ethical challenges of *in utero* gene therapy, *Nat. Genet.*, 2000, **24**, 107.
[11] T. W. Sadler, *Langman's Medical Embryology*, Lippincott, Williams and Wilkins, Baltimore, 2004.
[12] T. Hollon, Researchers and regulators reflect on first gene therapy death, *Nat. Med.*, 2000, **6**, 6.
[13] J. Kimmelman, Recent developments in gene transfer: Risks and ethics, *B M J*, 2005, **330**, 79–82.
[14] C. E. Thomas, A. Ehrhardt and M. A. Kay, Progress and problems with the use of viral vectors for gene therapy, *Nat. Rev. Genet.*, 2003, **4**, 346–358.
[15] E. R. Maher, Imprinting and assisted reproductive technology, *Hum. Mol. Genet.*, 2005, **14**, R133–R138.
[16] R. M. Winston and K. Hardy, Are we ignoring potential dangers of *in vitro* fertilization and related treatments? *Nat. Cell Biol.*, 2002, **4**, s14–s18.
[17] C. Allegrucci, A. Thurston, E. Lucas and L. Young, Epigenetics and the germline, *Reproduction*, 2005, **129**, 37–149.

[18] J. A. Robertson, Procreative liberty in the era of genomics, *Am. J. Law Med.*, 2003, **29**, 439–487.
[19] J. A. Robertson, Procreative liberty and harm to offspring in assisted reproduction, *Am. J. Law Med.*, 2004, **30**, 7–40.
[20] L. Nordenfelt, *Action, Ability and Health. Essays in the Philosophy of Action and Welfare.* Kluwer, Dordrecht, 2000.
[21] K. Zeiler, Reproductive autonomous choice – a cherished illusion? Reproductive autonomy examined in the context of preimplantation genetic diagnosis, *Med. Health Care Phil.*, 2004, **7**, 175–183.
[22] W. van den Daele, The spectre of coercion: is public health genetics the route to policies of enforced disease prevention? Community Genet., 2006, **9**, 40–49.
[23] I. D. Young, *Introduction to Risk Calculation in Genetic Counseling*, 2nd edn, Oxford University Press, Oxford, NY, 1999.
[24] H. Jonas, *The Imperative of Responsibility: In Search of an Ethics for the Technological Age*, University of Chicago Press, Chicago, London, 1984.
[25] H. H. Kazazian, Mobile elements: Drivers of genome evolution, *Science*, 2004, **303**, 1626–1632.
[26] H. H. Kazazian, An estimated frequency of endogenous insertional mutations in humans, *Nat. Genet.*, 1999, **22**, 130.
[27] K. Senior, Germline gene transfer during gene therapy: Reassessing the risks, *Mol. Med. Today*, 1999, **5**, 371.
[28] J. W. Gordon, Germline alteration by gene therapy: Assessing and reducing the risks, *Mol. Med. Today*, 1998, **4**, 468–470.
[29] N. M. King, Accident and desire: Inadvertent germline effects in clinical research, *Hastings Cent. Rep.*, 2003, **33**, 25–30.
[30] K. A. High, The risks of germline gene transfer, *Hastings Cent. Rep.*, 2003, **33**, 3.
[31] A. Eckhardt, Gentherapie, Schweizerischer Wissenschaftsrat: Programme Technology Assessment, TA 32/1999, 1999.
[32] National Institutes of Health, NIH guidelines for research involving recombinant DNA molecules, 2002. http://www4.od.nih.gov/oba/rac/guidelines/guidelines.html
[33] R. Dresser, Designing babies: Human research issues, *IRB: Ethics Hum. Res.*, 2004, **26**, 1–8.
[34] D. Voss, *Rechtsfragen der Keimbahntherapie*. Kovač, Hamburg, 2001.
[35] Act for Protection of Embryos (The Embryo Protection Act – EPA) issued in Bonn, Federal Law Gazette, Part I, No. 69: 2746, 19th December 1990. http://www.bundestag.de/parlament/kommissionen/ethik_med/archiv/embryonenschutzgesetz_engl.pdf
[36] D. Resnik, P. J. Langer and H. B. Steinkraus, *Human Germline Gene Therapy: Scientific, Moral and Political Issues*, R.G. Landes, Austin, 1999.
[37] K. R. Smith, Gene therapy: Theoretical and bioethical concepts, *Arch. Med. Res.*, 2003, **34**, 247–268.
[38] M. K. Grün and B. Morsey, *Prospektive Gesetzesfolgenabschätzung zum Problembereich somatische Gentherapie*, Forschungsinstitut für öffentliche Verwaltung, Speyer, 1997.

"Ghost of Christmas Past" ... Eugenics and other Moral Dilemmas Surrounding Genetic Interventions: A Discussion in the Context of Virtue Ethics

Agomoni Ganguli

Ethics Centre, University of Zurich, Switzerland

Abstract

It is never an easy task to talk about the moral dilemmas surrounding a technology that is still far from becoming an established medical procedure. One always runs the risk of either stating the obvious pitfalls of medical technologies or being tempted to speculate and be dismissed as having too wild an imagination. However, reproductive technologies, even at their embryonic stages, continue to stroke our hopes and fears and it is worth discussing them even when we have an unclear picture of what the future holds.

In the first part, this paper looks at the history of Eugenics and its influence on current debate surrounding gene therapy and other genetic technologies. Then, pointing out the blurred distinction between therapy and enhancement, it goes on to discuss the various moral dilemmas in the context of virtue ethics.

Keywords: reproductive technologies, eugenics, gene therapy, enhancement, virtue ethics, virtues.

Contents

1. Introduction 124
2. The ghost of christmas past 125
3. The ghost of christmas present 126
4. Introducing virtue ethics 129
   4.1. Virtues 129
   4.2. Eudaimonia and the role of narrative 129
   4.3. Emotions and special relationships 130
   4.4. Providing a rich vocabulary 130
   4.5. Character, motive and action 131
   4.6. Phronesis and the golden mean: further guidance on action 131
5. Virtue ethics, therapy and enhancement 132
   5.1. Tackling the therapy/enhancement distinction 133
   5.2. The role of parents 133
   5.3. Parents are citizens 138
   5.4. The role of doctors and scientists 139
6. Conclusions 140
References 141

1. INTRODUCTION

It is never an easy task to talk about the moral dilemmas surrounding a technology that is still far from becoming an established medical procedure. One always runs the risk of either stating the obvious pitfalls of medical technologies or being tempted to speculate and be dismissed as having too wild an imagination. However, reproductive technologies, even at their embryonic stages, continue to stroke our hopes and fears and it is worth discussing them even when we have an unclear picture of what the future holds.

In *A Christmas Carol*, Dickens tells the story of an old, miserly businessman named Scrooge who, on Christmas Eve, is visited by three ghosts. They appear in order to show him the error of his ways and to become a reformed man, thus avoiding the terrible fate awaiting him. Similarly, whenever we discuss reproductive technologies, there lures in the background the ghost of eugenics, a memory of past misdeeds and errors that seems to warn us about an over-enthusiastic approach and a lack of perception that might lead to tragic consequences. Eugenics has, even after its apparent death, left its footprint on the medical, historical and political debates in Europe.

"Bah! Humbug!" says the Scrooge in us, bury it deep inside your collective conscience and move on! But like Scrooge's ghosts, the memory of eugenics is present in all debates, rattling its chains and shaking us until we are reminded of our dark past. Gene therapy, like most new biomedical technologies, cannot escape this fate. Once again, we turn to face our fears. What will the ghost of eugenics tell us about the future of genetic interventions? Are we heading towards a future even darker than our past? How can ethics guide our actions so that history, in this case, does not repeat itself?

In one of his discussions on gene therapy, the philosopher John Harris, quotes the following definition:

Eugenics: A. adj. Pertaining or adapted to the production of fine offspring. B. sb. in pl. The science which treats of this.

It is indeed safer to start the discussion with a definition because the term "eugenics" has been used in so many contexts that one can easily get confused as to what exactly is being discussed. The following discussion is based on the above definition. Furthermore, I will use the noun "Eugenics" to refer to the science of Eugenics as it was known in the early 20th century and to all of its associated historical and political connotation, and the adjective "eugenic" to refer to current procedures or actions that relate to the "production of fine offspring". This is with the hope that current reproductive technologies, including gene therapy, can be judged on their own merits rather than on their association with history.

2. THE GHOST OF CHRISTMAS PAST

Eugenics first came to light as an ideology concerned with improving the health and moral constitution of mankind. Based on a thoroughly incomplete knowledge about human heredity, a lack of respect for the individual, and driven forward by economic and social pressures, it became the cause of some of the worst forms of coercion and atrocities known in history.

It was in the 1880s that the British scholar Galton first coined the term "eugenics", inspired from the Greek word meaning "noble in heredity". He believed that "what nature does blindly, slowly and ruthlessly, man may do providently, quickly and kindly" (Galton as cited by Kevles [1]). Soon, his ideas became popular among scientists, social workers, government officials and some of the greatest thinkers of the 20th century, including Nietzsche and Shaw [2]. With a very basic knowledge of Mendelian inheritance, the proponents of Eugenics put forward the theory that feeblemindedness, epilepsy, criminality, insanity, alcoholism, pauperism and many other such traits ran into families and were inherited in exactly the same way as coat colour in guinea pigs [3]. Educational and social welfare programmes began to rank people in terms of genetic "value", advocating both positive and negative eugenic measures. The former, to encourage those with "superior" genes to marry well and found large families, and the latter to stop those with "inferior" genes from spreading them, including with the help of sterilization. By 1920, 24 US states had eugenic sterilization laws, most of which were compulsory for mentally defective people [3]. In Sweden, between 1934 and 1975, over 60,000 individuals were sterilized legally. In an interview many years later, one of the doctors who had carried out such measures said "we dreamed that we could improve human body and soul ... that's how the geneticists saw it then and some of them see it still today" [4].

It is worth noting that although Eugenics is often associated with Nazism, it did not originate with the Third Reich. The Nazi merely embraced some of its features in order to better implement their political and social ideologies. Müller-Hill points out that the case of Eugenics in Germany is so untypical that it might not even be a pertinent example. In fact, Fischer, a geneticist and the first director of Kaiser Wilhelm-Gesellschaft received funds from the Rockefeller Foundation to study Eugenics well before Hitler came to power [4]. In other words, during the early days, much of what was happening in Germany was no more than an echo than what was taking place elsewhere.

There was enough opposition during the Weimar Republic to stop the establishment of racial hygiene laws. However, things changed quickly in the Third Reich. Weiss writes that "[i]n view of the institutional development and increasing popularity of eugenics during the 1920s and the early years of the Great Depression, it is not surprising that many racial hygienists welcomed the Nazi takeover as an opportunity to see eugenic measures, such as a compulsory mass sterilization law implemented" [5]. Apart from sterilization, other measures included what was euphemistically known as euthanasia: the termination of unworthy life (Vernichtung lebensunwerten Lebens). Disabled children and adults were killed in hospitals through starvation and overdoses of barbiturates [6]. During the Third Reich, eugenic measures (mainly sterilization and euthanasia) were implemented on mentally retarded children and adults, homosexuals, the Sinti and Roma and on individuals of Jewish descent. After the Second World War, the popularity of Eugenics greatly diminished, although some practices, including selective immigration laws and sterilization, existed well into the 1970s.

3. THE GHOST OF CHRISTMAS PRESENT

Given this sinister backdrop, it is no wonder that the ghost of Eugenics knocks at our door every time we discuss techniques that might intervene in human reproduction. However, it is worth looking at the characteristics of Eugenics which revive this fear and are worthy of ethical consideration. A prominent feature of Eugenics was coercion and the restriction of reproductive freedom. Coercion and restriction are not intrinsic components of reproductive technologies. Rather, they are particular results of how a group, community or government uses these technologies. One might argue that human rights being such a strong component of today's democracies, it is difficult to imagine that a government would be able to resort to the practice of Eugenics. Indeed, what people's morality does not accept today, few governments can impose. In the same way, given the ease with which we can cross borders and pay for a certain technology that is not available at home, few governments can restrict what our morality accepts. Indeed, the restrictions that currently exist in reproductive technologies have given rise to various forms of reproductive tourism [7]. Eugenics, where it exists today, is a reflection of individual choices rather than of governmental policies. An example of current eugenic practices would be the accepted use of abortion when pre-natal diagnosis reveals a foetal abnormality. An even more pertinent example would be that of genetic testing in some Ashkenazi Jewish communities, where carrier screening has reduced the occurrence of Tay-Sachs, an autosomal recessive neuro-degenerative disorder, by 90% [8]. This seems to be mainly the result of the work of an organization known as Dor Yeshorim (generation of the righteous), founded by a Rabbi who lost four of his children to Tay-Sachs. In close-knit Ashkenazi communities in the US, Europe and Israel, children under this programme are routinely tested for Tay-Sachs and their results (although undisclosed to them) are registered on a database. Because most marriages in these communities are arranged, one of the parties concerned consults the database and finds out whether the two young people are genetically compatible. Two carriers of Tay-Sachs would be advised against such a match. Although some concerns regarding the stigmatization of carriers have been raised, this eugenic method continues to be used with the support of the members of the community.[1]

One might argue then that if people are to restrict reproductive freedom or carry out eugenic measures, they do not need gene therapy. In fact, it is sometimes argued that even the simple choice of a partner is to a certain extent eugenic. As mentioned earlier, it is not the intrinsic eugenic nature of a technique that should worry us but the use it might be put to. Yet a particular fear remains concerning gene therapy or any genetic intervention. A particular worry is that somatic therapy will eventually lead to targeted germline intervention. In the Tay-Sachs example above, the gene continues to be carried through the generations; there is no question of changing the gene pool, only of a procedure to avoid the occurrence of the disorder. Germline therapy would imply eliminating a certain genetic mutation from the gene pool altogether, and as such, echoes the arrogance and hubris of early Eugenics and its tendencies to "improve the stock". If genetic interventions continue to be costly procedures, there may soon be an economic pressure to encourage a permanent measure rather than one that has to be carried out for each generation, *in utero* or at a stage before implantation. This perhaps highlights one of the most worrying features of germline intervention: its irreversibility. We would be making a choice that would affect all future generations and no doubt this would be a eugenic measure. Seen from this point, permanently engineering a gene out of the genetic make-up of human beings can indeed seem a frightening prospect. However, one might argue that irreversibility can also be a good thing. Notwithstanding it being a eugenic procedure, would it really be worse to eradicate Tay-Sachs or Huntington's than it was to eradicate small pox? One may argue that the difference is ethically irrelevant.

Having said that science is still at a stage where we cannot possibly imagine all the consequences of removing or silencing a certain mutation. To enthusiastically pursue such a radical measure would be as foolish as the undertakings of early Eugenicists. However, if in the future, we could show that germline therapy is safe, it would be difficult to find sound arguments to prevent it based solely on its irreversibility. Also, we could imagine that if we can engineer a gene or mutation out, theoretically, we should also be able to engineer it back in, should the need arise.

Even if we come to accept the irreversibility of germline therapy, we are still faced with other problems, such as the issue of subtle coercion. In the early 20th century, eugenic measures gained popularity due to two main

---

[1] Source of information: *http://www.shidduchim.info/medical.html#DorYeshorim*. Reliability of the information is unclear.

factors: a narrow conception of health and normalcy as well as economic pressures. It is not impossible that as gene therapy becomes an established procedure, the economic and social pressures might "force" parents to make use of available techniques. Some argue that encouraging people to only have healthy children devalues the life of those who are born with a disability [9]. If this is true, it would indeed echo the arrogance and intolerance of early Eugenics. But is it true? In order to look at this question, we have to first establish that health is an important good. Callahan argues that "if health is of no value then what is wrong with letting a person languish in pain on the street with a tin box when a prosthetic leg or a seeing-eye dog could make them independent"? [10]. Unlike the reproductive technologies available now, gene therapy offers the possibility to have a healthy child without having to consider the option of abortion. *In utero* gene therapy could further avoid the creation of embryos, most of which would then have to be discarded. In other words, with gene therapy one does not face the moral dilemma of either giving birth to a child with a disability or terminating a pregnancy. It offers parents the possibility to have a child without the disability. In this case, it is only the disability that is devalued, not the child in question nor the procreative freedom of its parents. One might argue that social or economic pressures are not the same oppressive tools they were in the days of Eugenics. There is no question that society needs to be more tolerant towards individuals with disability. Much progress is needed in that domain. However, encouraging tolerance does not need to impede medical progress. As Callahan says, if Huntington's were a bacterial disease, would we not straight away treat it with antibiotics? [10]. Moreover, we would probably hope that parents would feel a certain obligation to treat their children with the proper medication in this case, just as they are expected to for other illnesses.

Only negative Eugenics as well its restrictive and coercive features have been dealt with up to this point. A more difficult topic in this case is positive Eugenics. As we have seen earlier, Eugenecists not only discouraged the spread of "inferior" genes but also encouraged the spread of "superior" genes and traits. However, gene therapy and more generally genetic interventions would provide us with a tool that early Eugenecists could not have dreamt of: the possibility to play with the genetic make-up of individuals to enhance their physical, mental and moral traits. This discussion is made even more complicated by the fact that it not always easy to differentiate between therapy and enhancement. There have been many discussions concerning therapy and enhancement in ethics. Most discussions in ethics follow either of two main schools of thoughts: Deontology or Utilitarianism. In this discussion, I would like to look at the debate on positive Eugenics and enhancement from the perspective of another school, that of virtue ethics. Virtue ethics, as the name suggests, is concerned not only with our actions and their moral values, but with the motivation, character and virtues that guide them. As mentioned earlier, policies and regulations cannot always restrict people's choices. It is therefore important that the ethical discussion should also focus with the personal ethics of the individuals concerned.

4. INTRODUCING VIRTUE ETHICS

The origins of virtue ethics can be traced back to the Greek thinkers such as Socrates and Plato but its main proponent was Aristotle. Of course, some of what was discussed then has lost its relevance in our modern days. However, much of its essence has relevance in current debates and can be found in the writings of contemporary virtue ethicists such as Hursthouse, Foot, Slote, Nussbaum and McIntyre.

4.1. Virtues

Virtue theory claims that human behaviour and actions are in accordance with morality if they are in accordance with the virtues. Virtues are described as good or admirable traits of character, expressed in action, desire, attitude, thought and reasoning [11]. They include compassion, courage, honesty, patience and other such traits. Virtues are refined through training, education and the moulding of one's character and desires in the correct way. A virtuous person, according to Aristotle, will fare better in life and ultimately attain the highest good, which he calls eudaimonia. Some authors see virtues as a state of character: if a person is virtuous, exercising the virtues comes naturally and easily to her. Others see the virtues as dispositions that allow one to become virtuous through performing virtuous acts, especially in demanding situations.

4.2. Eudaimonia and the role of narrative

The Greek term eudaimonia has no one-word translation in English. Hursthouse describes it as a combination of well-being, happiness and flourishing [11]. It is the ultimate goal of a human being, a part of her function. The virtues are therefore human needs that benefit their possessor. McIntyre, who describes human life as a whole with a history which links our birth, life and death in a continuous narrative, describes the virtues as dispositions, which will enable us to live through this narrative sustaining us in the relevant kind of quest for the good, by helping us to "overcome the harms, dangers, temptations and distractions which we encounter, and which will furnish us with increasing self-knowledge and increasing knowledge of the good" [12]. The idea of a narrative is of particular importance because, especially in the context of ethical dilemmas, our lives lead us to assume many roles, each requiring the exercise of various virtues.

4.3. Emotions and special relationships

E. M. Forster once remarked that if it came to a choice between betraying his country and betraying his friend, he hoped that he would have the courage to betray his country [13]. Virtue ethics recognize the importance of emotions and special relationships, even within the framework of moral theory. It points out the importance of human instinct, reminding us that morality can go beyond the adherence to principles and can also draw its strength from deep human feelings. Moreover, it recognizes that the ties we have with some people due to our emotion, can also have distinctive moral significance.

In a passage in *Wonderwoman and Superman*, Harris writes: "If your child is suffering from a fatal kidney disease and mine has the only available kidney, would I be wrong to risk my child to save yours? I do not think that I would in fact do so, but that is because I am a parent and I have fierce protective feelings towards my child" (Harris as quoted by Hayry [14]). Harris does not see the need for any further explanation because the feelings involved are beyond reasoning or ethical justification. This is a case that Utilitarians or Deontologists would say does not belong to ethical reasoning or justification. Kant, for example would say that where such powerful emotions are involved, we could be trusted to make the right decisions. Principles are needed where emotions are not involved. However, our emotions and feelings often have a very important impact on our power of deliberation over moral issues and as such cannot and should not be separated from reasoning and justification.

Hursthouse uses the example of jumping into the river to save one's child. To say "this is my duty or the right thing to do" seems repellently self-righteous. The right reason is "she is my child". However, as the source moral thinking, human emotion should play a role in the discussion of ethics and morality. Furthermore, actions judged according to the rules of morality are not independent of motives and motives are strongly influenced by emotions.

4.4. Providing a rich vocabulary

An interesting aspect of virtue ethics is its use of aretaic concepts: the use of words such as "admirable", "honest", "fair", "generous" and "callous", as opposed to the usual terms used in Deontology or Utilitarianism, such as "right" and "wrong". Because virtue ethics uses various words to describe a person's character, motivation and emotions, it provides a rich vocabulary that is somewhat lacking in those theories that concentrate mainly on actions. As we shall see later, an action can be more than just right or wrong. If a parent's decision to enhance a child promotes racism, then that act can be both benevolent for that child (who himself will not suffer from racism) and callous (because it promotes racism). Narrowing the discussion at the start on such a case to either "right" or "wrong" might fail to provoke our imagination in creating a complete picture of the situation. Of course, it is possible then to derive the "rightness" or "wrongness" of the action. However that would be the conclusion of our reasoning, rather than the building blocks of it.

It seems that we often ignore how strongly (sometimes dangerously) words shape debates in biomedical ethics. Where there are images of "designing children", "genetic control", "playing God", "tinkering with genes", "threatening our humanity" and "eugenics", deontic words such as good or bad, right or wrong may sound weak. Words taken from the realms of ethics such as ignoble, admirable, arrogant, conceited, tolerant, moderate, loyal, to go with such images, words which in turn can illustrate the motives, characters, put into perspective the ethical credibility of the individuals and the actions in question.

4.5. Character, motive and action

As we are beginning to see, the main question raised in virtue ethics is not "what should I do but what kind of a person should I be?" Character and motives are often reflected in actions. As Bernard May points out "It is obvious that a man cannot just be; he can be what he is by doing what he does; his moral qualities are ascribed to him because of his actions, which are said to manifest those qualities" [15]. Character and the correct motives are built through constant education, training and the self-questioning. A person with a sound character and confidence in her motives can rely more strongly on her intuitions and her reasoning while taking decisions. Mastering the virtues gives her the ability to perceive the world more clearly. However, owing to the importance it gives to motives and character rather than action, virtue ethics is often criticized as saying that it does not matter what is being done as long as the motive is right. This seems to be a misunderstanding of the concept of virtue. The correct grasp and mastery of a virtue include being able to use it in the right place and to the right extent. Let us take the example of courage: the true expression of courage is in the face of danger, showing it off where there is no danger is a sign of arrogance and sometimes even cowardice and people who do so are often the ones who flee in the face of real danger. A truly virtuous person has the ability to recognize which virtue is called for, and to what extent it should be used.

4.6. Phronesis and the golden mean: Further guidance on action

Two other considerations can guide a person's actions. The first is the idea of the mean. According to Aristotle, a virtue in its correct state is a mean between an excess and a deficiency [16]. If we go back to the example of courage, the correct expression of courage lies mid-way between cowardice and foolhardiness. The same applies to, for instance, anger. Too much is not good, nor is too little. But we must note that virtue ethics is not encouraging moderation in all circumstances. Some situations require more anger, and some less. The same amount of courage might seem foolhardy in some situations, cowardly in others. Where the mean lies depends on the situation.

This brings us to the idea of phronesis. Phronesis, also known as practical wisdom, is what enables us to perceive a situation, deliberate correctly and use the right virtue to the right extent. Nussbaum defines it as: judgement, discernment, experience, intelligence and a sign of well-trained desires and emotions [17]. This practical wisdom comes with time and training and is finely tuned by experience. Phronesis plays perhaps, one of the most important roles in virtue theory, especially in the context of role in applied ethics.

5. VIRTUE ETHICS, THERAPY AND ENHANCEMENT

Before looking at how virtue ethics would deal with the question of therapy versus enhancement, let us have a quick overview of how it is discussed in the current literature. Many experts draw a distinction between therapy and enhancement, followed by the claim that while the former is health related and therefore obviously morally acceptable, the latter is a sort of cosmetic intervention, which one can forego. The philosopher Eric Juengst suggests that therapy can be distinguished from enhancement by its objective to treat or prevent disease [18]. At first glance this seems quite plausible. The telos of medicine is to reduce suffering and cure when possible. It does not include fantasies such as skin that glows in the dark. Now, the problem is defining a "disease" or a "disability" on the one hand and "normalcy" on the other. We do not actually have an objective standard against which to measure specific conditions. "Normal" usually refers to some frequency of a characteristic in a population. We measure everyone's height and call the average height "normal" [18]. Another way to define normal is what Norman Daniels calls "species-typical functioning": normalcy is whatever falls "within the limits set by natural endowments" [19].

Now, as is sometimes pointed out, it all depends on how we measure and evaluate the species and we come up against obvious problem when trying to define psychological and behavioural disorders [20]. A normal or minimum standard might differ tremendously from one group to another. "De rigueur in some circles are private music lessons and trips to Europe, whereas in other providing eight years of schooling is a major accomplishment [21].

In the case of ageing, for example, genetic manipulations could be considered both an enhancement (by increasing life expectancy) and a therapy against age-related illnesses [22]. Let us also look at strengthening the immunity system. An intervention to boost the immunity systems of those of us not suffering from any major illnesses would be considered an "enhancement". Yet, how could we explain not protecting ourselves from the everincreasing pollution in the environment? It is hardly different from an MMR shot or a polio vaccine. This is the line of thought that Harris follows and indeed, in his article, goes on to say that just as genetic therapy should be encouraged, so should genetic enhancement in certain cases [23].

5.1. Tackling the therapy/enhancement distinction

Let us come now to the main question. If so many thinkers cannot come to a satisfactory differentiation between therapy and enhancement and establish two definitions that would draw a simple line between what is permissible and what is not, can virtue ethics find a satisfying answer? A straightforward answer would be: no, it cannot. What virtue ethics can do, however, is to qualify each case in detail, draw a precise picture of the individuals involved, of the motives and emotions surrounding the case, and highlight the ethical points that need to be considered. This is an approach used by Hursthouse in her discussion on abortion. In her opinion, facts about the status of the foetus or women's right are irrelevant. Pregnancy and parenthood are extremely important events in life; the ethical dilemma should be solved by looking at each case, analysing the motives and situation of the mother and the subsequent love and care that the child will receive. Similarly, in this discussion, it is pointless to go try to reach a consensus on what is generally permissible and what is not. Virtue ethics does not need to say that therapy is right and enhancement, or positive eugenic measures are necessarily wrong. It can afford to say that all genetic intervention aimed at improvement (be it health or natural talents) could be defined as an enhancement and some of these can be within the bounds of morality. In order to reach the correct deliberation regarding enhancement, we need to look at the character and motives of those involved, the emotions and considerations that will push someone to choose enhancement and the virtues and vices surrounding each choice. It is of course impossible to establish a thought experiment for each theoretical possibility, but it is possible to highlight the main issues with the help of a few important considerations.

5.2. The role of parents

The three main actors in our present discussion are the child to be enhanced, the parents making the choice and the doctors/scientists who will provide the technologies and carry out the interventions. In early Eugenics, a fourth actor was the State. Today perhaps, the fourth actor is society, with its perceptions, concepts and expectations as well as the support it provides for persons with disabilities.

Buchanan and his co-workers rightly point out that parents are generally regarded as having the permission, and some would say an obligation to produce the "best" children they can [24]. Indeed, parents are expected to give children the best education within their means, some moral guidance and to look after their health and general safety. Savulescu takes this idea a step further and says that parents should use reproductive technologies where available to produce the "best" children they can as a part of what he calls "procreative beneficence". He even asserts that where there is a social preference for one of the two sexes, parents should avail the technologies to choose the sex which fares better, for the sake of the future child [24]. There is no doubt that this attitude reflects eugenic tendencies, albeit not Eugenics as we have known it. We shall look at the idea of social preferences and the role of parents in the social context later on in the discussion.

Here, it is perhaps interesting to look back at McIntyre's idea of a narrative. To some, parenthood is a part of their narrative; others have a vocation or an aim in life, which does not include becoming a parent. Those of us who do plan to become parents are aware of taking up an important step in life. If a young woman has a child, many of her actions and desires will be influenced by the fact that she is a mother and she will be expected to exercise the virtues related to motherhood such as care, benevolence, patience, dependability and even a certain amount of sagacity. If a couple decides to have children in order to compete with the neighbours or as part of their "life-styles", just as they might choose furniture to express their personality [25], it would be an expression of light-minded, over-indulgent and selfish characters. With or without the use of genetic enhancement, the action of these parents would be morally unacceptable.

Thankfully, most parents are not like that. They would like to have children so they may enjoy the wonders of parenthood and bring up a child with love and care. However, even such parents should be very careful about the steps they take. As McGee points out: "because parenting is subtle, sophisticated and enormously complicated, it is not at all surprizing that we should be unaware of our own motivations" [20]. In many ways, a parent is the safest judge because her motivations when choosing enhancements are not tainted by Utilitarian calculations. However, after taking all the practical facts into account, the best course can only be decided once the parent has questioned herself on the motives that have influenced her choice. Of course, her choices will be always bound by the norms and the tolerance of the society in which she lives.

Stock says, "Some parents insist that their children study hard and earn good grades. Some push their kids towards sports. Some want outgoing and popular offspring. Whether we guide our children with a heavy hand or are subtle and indirect, the paths we try to choose for them often tell more about us than about them or who they will become" [22]. In enhancing traits, parents have to decide whether they are truly taking this step in order to offer the best life to their child or whether they are gratifying their own desires. Giving a child piano lessons when the child has shown an inclination is one thing, highly enhancing her musical talent before birth so that she could become a musical genius might be a sign of vanity, excessive pride and a lack of concern for the child's other talents. This is what McGee calls "the sin of calculativeness" [20]. The parent, obsessed by her own expectations, might create an atmosphere of harsh training and expectation that would instil a certain aggressiveness and a sense of competition in the child, rather than a real love for music. Moreover, such constraint might impede a child from fulfilling a life narrative that would satisfy her quest for flourishing or eudaimonia. Again, it is worth noting here that it is not the means of enhancement that is the most important. Environmental approaches can have similar effects to genetic ones.

The Greek term "eudaimonia", used by Aristotle is difficult for us to grasp, both in linguistic and conceptual terms. McIntyre says, "what constitutes the good for man is a complete human life lived at its best" [13]. In other words, eudaimonia is the state attained by an individual living her life within the moral boundaries set by the virtues and fulfilling her roles and attaining her goals in life. As Trilling points out "the sense or significance of our lives depends on how we live them" and that our lives consist of "planned undertakings, which, to a large extent, we control and for which we are responsible" [26]. An important point in this discussion is that in order to fulfil our roles as human beings there is a need for a life plan, a sort of telos, that will guide our actions and our motives in the right direction. In the context of enhancement, the fulfilment of a lifeplan will not conflict with the choices of somatic enhancement in an adult. "If we know the telos of our life, we should be able to judge, with greater or lesser degrees of specificity the kinds of enhancements that would further our moral formation in light of that telos and the kinds that would detract from it" [27]. However, when children are enhanced before birth, it is not them but their parents who are defining their telos and life plans.

Plato suggested that each man should do what he is most fitted for. However, that is very different from designing someone to fulfil a preordained role [28].

Many fathers with a successful career have wanted their eldest son to follow their footstep and many eldest sons take painstaking measures to avoid following their father's footsteps. Human beings' interests, goals and aspirations, change enormously as they grow older and according to the environment they grow up in. Part of the magic of life lies in our ability to explore and experience various aspects of the world, before finding our own path through life. Agar calls lifeplans both environmentally specific and environmentally sensitive [29]. Specificity comes from what is available to you as you grow up and sensitivity is due to the fact that even small changes in the environment (a television programme, an inspiring talk) can trigger huge changes in one's life plans. While sensitivity opens many a new doors to us, specificity can act as a constraint. We might think that enhancement will reinforce talents and therefore open many doors but, because of its irreversibility, it can also act as a severe constraint. Every person has to choose her own narrative and our role as parents would be to leave as many doors open as possible.

Unfortunately, all cases will not be that simple. In 2000, Reindal challenges Harris' statement that it is morally wrong to produce children who will be significantly harmed because of their constitution [30]. Reindal gives the example of individuals with deafness, explaining that such parents could very well decide to choose deafness in their children. They would do so because they believe that in this way, they can relate more easily to their child, communicate better with her, help her integrate into their community and the deaf culture; in a few words: to be better parents to her and thus give her a better life. As Stock points out, this might distress us but that no one can really argue that these parents are injuring a healthy child, they would simply be choosing deafness [22]. However, we have to remember that this is not entirely comparable to a cultural background. People from various cultures and are able to lead similar lives to everyone else while still being true to their culture. When choosing deafness for her child, a parent has to ask herself what her real motives are. Is she really making this decision so she can relate better to her child, so that her child has an easier and better life, in other words be a better parent? Or is she choosing it so that her child can relate better to her, understand her better, be a better child to her? In a world where most people are not deaf, a world where her child would most probably need to communicate with the rest of the world, where more doors would be open to her were she not deaf, would it not be egotistic to deliberately choose deafness? The first duty of a parent is to support her child and some find themselves in more difficult situations than others in raising children. However, the life of a child does not start and finish within the family and the community. These are only a support for a path the child chooses as an adult.

This becomes even more obvious if we compare deafness to other cultural backgrounds. In a multicultural society like ours, many of us come from very different traditions and cultures. Parents naturally want that their children should feel just as comfortable in their culture at home as they do in the culture where they have grown up, a culture they share with friends and colleagues. While most people find it admirable that a child can be at ease in two cultures, what would we think of parents who specifically programmed their child to learn their mother tongue only, have an ear only for their traditional music and could taste nothing but their mothers' recipes? We would probably think the parents utterly selfish and ignorant, if not a bit mad. Can we denounce as fundamentalists those who will hear nothing but the words of their own religion and yet condone deaf parents' choice as simple preference instead of denouncing it as irresponsible? If such parents were responsible, prudent and aware of their critical role as parents, they might want to teach their child the sign language and bring her up within the community so that she would develop an affinity for her surroundings and a bond with her parents and other members of the community, but not constrain her within the community and force her to share their disability.

Finally, as far as motives and virtues are concerned, parents over-enthusiastic for enhancements might see certain virtues lost in their children. Enhancing a child to excel in say sportive, musical or intellectual competitions will also instil certain values in her. That child can easily grow to become determined, hardworking, ambitious but also callous and malevolent, even dishonest in order to reach the top. Moreover, having a talent only for achievement is not always a blessing. Such people might not be able to take defeat with a certain good humour. Stock, although an advocate for enhancement, nonetheless points out that: "None of us wishes to see our children suffer, but if we could protect them from all the dangers and pains of life, we might ultimately diminish them by leaving them untested and shallow. After all, some of the influences that we, in hindsight, most value in our own lives are the painful failures and defeats we have overcome, the jarring potholes that have sent us lurching down new roads" [22].

Perhaps the obvious thing to do would be to enhance a person in a way that would allow her to follow any of the life plans she chooses. Most parents however, even with pure motives, will probably want to fine tune their children's intelligence (a useful trait in most cases), perhaps give them a couple of musical and sportive genes just in case the child shows an interest in either direction and probably throw in some kindness and generosity for good measure. It is more difficult to choose psychological and moral traits because it is difficult to decide whether a child would fare better as an introvert or an extrovert. Obvious targets would probably be the elimination of genes that cause genetic disorders, genes (if they exist) that pushes one towards antisocial behaviour and other aggressive traits such as excessive arrogance, conceit or greed. Parents making such choices would not be crossing the boundaries of morality. But, even in the cases where we can give "a little bit of everything", we will come across two problems. First, some children who are more or less good at everything in school often find it hard to choose a subject for higher studies and sometimes take longer than other to find their right aim in life. Genetic enhancement to help children excel in everything might only add to their confusion, leading to a sort of schizophrenic life plan. Second, having a little talent for everything might result in creating a "jack of all trades and master of none" where all enhanced children are good at everything but since they are not able to focus their attention, concentration and energy on one or a couple of talents they remain an average in all domains. The meaning of diverse life plans and hence of eudaimonia might slowly disappear. A good basketball player will never be a professional jockey and a boxer will never ice-skate at the Winter Olympics. And if they are both, it is likely that they are not very good at either.

Even if parents limit their choices to very safe enhancements such as intelligence and virtues, they might actually subdue some of the best talents in their children. Walters argues that when aggression is channelled into academic or athletic competition, within reasonable bounds, it can contribute in important ways to an individual's self-development. He gives the example of great revolutionaries and political figures such as Gandhi and Martin Luther King, who must have had very stubborn and quarrelsome personalities to have achieved their goals.

5.3. Parents are citizens

McIntyre looks at human life as a continuum where virtues are expressed in all actions and circumstances. It is true that a person's life cannot be compartmentalised in a way that she would exercise virtues in one compartment and not in the other. However, modern life is also an amalgamation of role plays. A mother might be entirely devoted to her child but she might also be a doctor, a teacher or a social worker, finding that similar virtues comes into play in different ways in her various roles in life. There is one other role that parents must always fulfil, apart from being parents: their roles as responsible members of a society. And as such, they are also responsible for shaping the views and norms of society. Most of the time, parents are able to fulfil both roles very well. However, there might be cases where parents would find that fulfilling the demands of one role might have a negative effect on the other. There may be cases when our strong feelings for our children might become a threat to our virtues as the members of a society. This is especially important in the context of enhancement because more often than not the values we place on certain traits such as cleverness, beauty and height, greatly depend on the dominating culture or concepts of a society. Even when our motives are honest and selfless, we sometimes need to consider the social values we are inculcating or reinforcing. This is similar to what Slote defines as a holistic approach to motivation: "Someone who cares about the well-being of one particular person and acts for that person's benefit acts, in some individualistic or narrow sense, from a good motive. But if such action involves neglecting the well-being of other people he knows or doesn't know, then it may demonstrate an overall bad character and may count as a wrongdoing" [31].

James Watson once said, rather infamously: "if you could find the gene which determines sexuality and a woman decides she doesn't want a homosexual child, well, let her (choose accordingly)" (Watson as quoted by Agar [29]). Although Watson makes it sound like a mere preference, it is not. If we can call this a preference, then we would also have to say that the "missing" baby girls in India and China are also due to a simple parental "preference". Village women in India do not drown their baby girls out of choice, they do it out of compulsion in already poor families, where raising a girl and marrying her off with a dowry would be tantamount to ruin. If they were given the option of a free sex change, they would probably consider it as an enhancement and would no doubt opt for it. In the Western, so-called "free world", a less tragic but perhaps just as pathetic scenario: women still continue to wasting much of their income, energy and natural beauty on cosmetic surgery in order to fit into the accepted standards of beauty. Not "fitting in" also brings a type of suffering. Many women would rather go through the pain and discomfort of surgery than feeling that they do not look their best. Eugenic tendencies exist in all cultures with or without the use of reproductive technologies.

This is perhaps one of the most delicate dilemmas facing genetic enhancement. A parent who has himself suffered much teasing as a child because of his height and has become shy and recluse as a result, finding it hard to communicate with colleagues as an adult even if he has not been teased for over 20 years, would perhaps not want to take the same risk with his child. It is understandable that as a good and protective parent, he would not want his child to be teased, even if the child, instead of growing shy like his father might grow into a tough boy. Unfortunately, such an attitude in a parent would merely reinforce the "heightist" values of our society. As Parens points out: "Even if we had no worries about how a given enhancement might undermine or violate the goals of medicine, we would still be concerned that it would promote complicity with harmful societal conception of normality". This is what Maggie Little calls the ethics of complicity. For an African American to want to look paler is not a whimsical preference [19]. It is due to the subtle pressure from society to conform to the norm.

Similarly, when parents decide to enhance their children, they should be aware that sometimes such enhancements rely on someone else's vulnerability and oppression. Only when women are not oppressed anywhere in the world, when plastic surgery is only used on accident victims and when our societies have ceased to be "heightist" or homophobic can we say that parents making such choices or enhancement are not sacrificing the virtues of fairness, compassion and justice. Otherwise as Cole-Turner points out, "one is complicitious when one endorses, promotes or unduly benefits from norms and practices that are morally suspect" [32]. However, sometimes people are forced to make unethical choices. This is what the women who carry out selective abortion for sex selection have to face. This was also what many health care workers faced when they had to carry out sterilization, euthanasia and other eugenic measures under an oppressive government. In these cases, it is useless to look only at individual ethics, the moral questions have to be dealt with from the perspective of society as a whole. The situation for health care workers is, however, slightly different when they work in democracies.

5.4. The role of doctors and scientists

It is perhaps the right time to say a few words about the motives and characters of doctors and scientists. Doctors, by virtue of their profession have a very important role in society. They are expected to be compassionate and trustworthy [33] as well as skilful and competent in their field. Owing to their prestige in the community, their statement can carry a great deal of weight [20]. As such, doctors must have the confidence that their deliberations and decisions are ethically sound. In order for enhancement to gain popularity, the demands of parents have to be met by professionals who are willing to research on and perform these enhancements, just as the measures used in early Eugenics had to be carried out by scientists, doctors, nurses and other health care workers.

A doctor's duty includes relieving her patient's suffering and attending to her health needs but her duty as a human being also includes not to encourage ostracism or to give in to complicity. Cole-Turner argues that a surgeon asked to make blacks look more "white" under apartheid should feel guilt [32]. In the same way, one wishes that more doctors and geneticists during the era of Eugenics had mustered the courage to protest. Conscientious and trustworthy doctors should be able to discern the implications and consequences of their choices and actions. Especially where money is involved, doctors should thoroughly check their motivation before they accept to endorse enhancement technologies. David Hyman says about cosmetic surgery that: "Medicine is debased and it becomes the handmaiden of vanity and self-indulgence, in the name of 'being your best'" (Hyman as quoted by Holtug, p.141 [34]). This is a crucial point because even when an enhancement is within the boundaries of the law, it is not necessarily within the duties and the virtues of medicine. As Slote argues, if a person harms others out of a sense of duty, such as a guard in a Nazi Prison camp, it is the reflection of inconsideration and as such the overall motivational state is a bad one and his professional conscientiousness does not redeem his action [31]. Therefore even when a doctor is doing his duty, that is, helping patience within the boundaries of the law, he might not be acting virtuously. Doctors are perhaps in a much better position to show equanimity and circumspection in the face of difficult choices because although they show compassion and benevolence towards their patients and their suffering, they can also show discernment and foresight in the face of dilemmas. Owing to their ability to be detached – unlike parents who are highly emotionally involved when it comes to their child – doctors are better able to take everything into consideration when advising parents, or indeed regulators and policy makers on their possible choices of enhancement. If most doctors recognize that certain enhancements or certain motives will be harmful for society, they need not comply with their patients' wishes and such enhancement will therefore rarely take place, if at all.

6. CONCLUSIONS

What then, has the ghost of Eugenics taught us about Christmas to come? While legislation and guidelines are needed to protect the rights and freedom of individuals, Eugenics perhaps shows us that simply "following the rules" may not always be the ethical path. History can help us to face our deepest fears and to learn from our past mistakes. However, our past does not have to convert us to biotechnological Luddites. Morality in general and the ethics of genetic interventions, regarding both therapy and enhancement, ultimately lie in the motives of those who are directly concerned: the parents, the scientists, the doctors and the rest of us choosing to embrace or selectively reject the possibilities created by biotechnology. As long as each of us feels directly responsible for the choices we make, we may indeed bury deep into our collective conscience the ghosts of past misdeeds.

REFERENCES

[1] D. J. Kevles, *In the Name of Eugenics*, University of California Press, Berkeley, 1985.
[2] K. Bayert, *Genethics, Technological Intervention in Human Reproduction as a Philosophical Problem*, Cambridge University Press, Cambridge, 1994.
[3] D. J. Kevles, International Eugenics, in *Deadly Medicine, Creating the Master Race* (ed. D. Kuntz), United States Holocaust Memorial Museum, Washington, D.C., 2004.
[4] B. Muller-Hill, Lessons from a dark and distant past, *Bioethics: An Anthology* (eds. H. Kuhse and P. Singer), Blackwell Publishers, Oxford, 2002, pp. 182–187.
[5] S. F. Weiss, German Eugenics, 1890–1933, *Deadly Medicine*, Creating the Master Race (ed. D. Kuntz), United States Holocaust Memorial Museum, Washington, D.C.,2004.
[6] H. Friedlander, Physicians as killers in Nazi Germany, *Medicine and Medical Ethics In Nazi Germany* (F. R. Nicosa and J. Huener), Berghahm Books, New York, 2002.
[7] G. Pennings, Reproductive tourism as moral pluralism in motion, *JME*, 2002, **28**, 337–341.
[8] R. Rozenberg and L. V. Pereirra, The frequency of Tay-Sachs disease causing mutations in the Brazilian Jewish population justifies a carrier screening program, *Sao Paulo Med. J.*, 2001, **119**(4), 146–149.
[9] M. Saxton, Why members of the disability community oppose prenatal diagnosis and selective abortion, in *Prenatal Testing and Disability Rights* (eds. E. Parens and A. Asch), Georgetown University Press, USA, 2000.
[10] J. C. Callahan, *Reproduction, Ethics and the Law*, Indiana University Press, USA, 1995.
[11] R. Hursthouse, *On Virtue Ethics*, Oxford University Press, Oxford, 1999.
[12] A. Macintyre Tradition and the virtues, in *Vice and Virtue in Everyday Life* (eds. Sommers and Sommers), Hartcourt Publishers, United States, 2001.
[13] A. Macintyre, *After Virtue: A Study in Moral Theory*, Duckworth, London, 1985.
[14] M. Hayry, *Playing God: Essays on Bioethics*, Helisinki University Press, Helisinki, 2001.
[15] B. May, Virtue or duty, in *Vice and Virtue in Everyday Life* (eds. C. H. Sommers and F. Sommers), Hartcourt Publishers, United States, 2001.
[16] P. Benn, *Ethics*, Routledge, United Kingdom, 2003.
[17] M. C. Nussbaum, *The Fragility of Goodness*, Cambridge University Press, Cambridge, 1999.
[18] M. J. Mehlman, *Wondergenes*, Indianapolis University Press, Indianapolis, 2003.
[19] E. Parens, Is Better Always Good? in *Enhancing Human Traits: Ethical and Social Implications*, Georgetown University Press, Washington, 1998.
[20] G. McGee, *The Perfect Baby*, Rowman & Littlefield, New York, 1997.

[21] L. M. Purdy, Genetics and reproductive risk, in *Bioethics* (eds. H. Kushe and P. Singer) Blackwell, Oxford, 2002.
[22] G. Stock, *Redesigning Humans*, Profile Books, London, 2002.
[23] J. Harris, Is gene therapy a form of Eugenics? in *Bioethics* (eds. H. Kushe and P. Singer) Blackwell, Oxford, 2002.
[24] A. Buchanan *et al.*, *From Chance to Choice*, Cambridge University Press, Cambridge, 2000.
[25] J. Savulescu, Procreative beneficence: Why we should select the best children, *Bioethics*, 2001, **15**(5), 165–170.
[26] H. Putnam, Cloning people, in *The Genetic Revolution and Human Rights*, The Oxford Amnesty Lectures (ed. J. Burley) Oxford University Press, England, 1998.
[27] L. Trilling and C. Taylor, Ethics of authenticity, in *Enhancing Human Traits: Ethical and Social Implications* (ed. Parens), Georgetown University Press, Washington, 1998.
[28] G. P. Mc Kenny, Enhancement and the ethical significance of vulnerability, in *Enhancing Human Traits: Ethical and Social Implications* (ed. E. Parens), Georgetown University Press, Washington, 1998.
[29] R. Chadwick, The perfect baby, in *Ethics, Reproduction and Genetic Control* (ed. R. Chadwick), Routledge, London, 1987.
[30] N. Agar, Liberal Eugenics, in *Bioethics* (eds. H. Kushe and P. Singer), Blackwell, Oxford, 2002.
[31] S. M. Reindal, Disability, gene therapy and Eugenics – A challenge to John Harris, *J. Med. Ethics*, 2000, **26**, 89–94.
[32] M. Slote, *Morals from Motives*, Oxford University Press, Oxford, 2001.
[33] R. Cole-Turner, Do means matter, in *Enhancing Human Traits: Ethical and Social Implications* (ed. Parens), Georgetown University Press, Washington, 1998.
[34] P. Gardiner, A virtue ethics approach to moral dilemmas in medicine, *J. Med. Ethics*, 2003, **29**, 297–302.
[35] N. Holtug, Does justice require genetic enhancements? *JME*, 1999, **25**, 137–143.

Biomedical Research and Ethical Regulations in China: Some Observations about Gene Therapy, Human Research, and Struggles of Interest

Ole Döring

Ruhr University Bochum Faculty for East Asian Studies China's History and Philosophy Building GB 1/137 D-44780 Bochum, Germany

Abstract

The intricacies of China's biopolitics and bioethics are frequently assessed through the lens of "cultural peculiarity" of China's moral common sense. Thereby it is overlooked that major factors of influence on contemporary legislation are at best indirectly construed in cultural terms, if not in conflict with traditional moral views. This paper introduces some recent trends in the ethical regulation of the biomedical sector in China and explains how they are inspired and particularly moulded by state political and stakeholders' interests. The licensing and industrial production of the world's first commercial gene therapy drug, "Gendicine", in China, will be discussed as a case example for the remarkable role researchers play in promoting the development of bioethics.

Keywords: confucianism, consent, informed, culture, ethics, bioethics, embryo, interest(s), morality, moral, social, standards, international, stem cells.

Contents

1. Introduction 144
2. A case of gene therapy – facts, figures and the emergence of policy 144
   2.1. Background state of the art 144
   2.2. On the market 145
   2.3. Clinical trials 147
   2.4. Regulations and success 149
3. On cultural expectations 150
   3.1. Culture misconceived 150
   3.2. China's approximation of ethical demarcations 151
      3.2.1. A chinese rubicon 152
      3.2.2. A chinese limes 152
4. The role of researchers in ethics regulations and the international state of the art: china's position 153
5. What it means for us? 156
References 157

1. INTRODUCTION

China's biopolitics and bioethics are frequently assessed through the lens of an assumed "cultural peculiarity" of China's moral common sense. This approach seems to respond to an attitude of respect for the differences and the underlying substantial identities of cultures. We hesitate to submit countries or societies, such as China, with an apparently alien cultural, historical or political background under norms and practices that have been crafted, for instance, in Europe or Northern America. This, among others, owes to considerations of Europe's imperialistic performance in history and the, more recently, resulting attempt to "learn from the past".

However, closer scrutiny indicates some relevant intricacies. For example, without proper reflection and information, one might overlook major relevant factors that influence contemporary legislation, which can rather be construed indirectly in cultural terms. Sometimes these factors may even stand in conflict with traditional moral views of a given society's culture.

Moreover, the meaning of culture and what it means to pay attention to or to respect cultural factors cannot be taken for granted, neither as a pure motive nor as a material key for the examination of the impact of a given culture.

Not at least, there is a danger that a cultural bias, of any brand, can be abused in merely instrumental discourse so that it supports the agenda of inexplicit interests. I shall attempt to show in the following how they can be misleading and stand in the way of understanding, for example, when they are built upon invented traditions or ungrounded generalizations.

The chosen strategy of this paper, that is, viewed from a critical cultural perspective [22], is to begin with an analysis of policy making in the areas of the life sciences and bioethics, as related to the establishing of universal (global) standards, and, in light of the reconstructed empirical situation, to undertake a preliminary cultural assessment of the context and content of particular areas of bioethics in China.

2. A CASE OF GENE THERAPY – FACTS, FIGURES AND THE EMERGENCE OF POLICY

2.1. Background state of the art

The modern Chinese biotechnology industry started two decades ago, in March 1986, with the government's 863 program. It was further developed with a focus on application of basic sciences in the National Basic Research Program (dubbed "973 Program"; of.: *http://www.973.gov.cn/English/Index.aspx*). It has now reached the turning point for a qualitative change: From "follow and copy" the Western countries to begin to innovate. China's accession into WTO has contributed to this change. Resolute adoption of international technical and procedural standards, in many areas, is helping China to pass the long way through the bottleneck of structural underdevelopment [1]. Landmark projects in Genomics, such as China's contribution to the Human Genome Project, the sequencing of model or strategic key genomes (notably rice), have confirmed this development. In the biopharmaceutical sector, alongside with the recombinant protein drugs now the world's first gene therapy drug is produced in China [23,24]. In the molecular diagnostics sector, various biochip products for clinical use have been brought to market by Chinese biotech companies [2].

2.2. On the market

In October 2003, according to Chinese and international response, Shenzhen SiBiono GeneTech made history. It became the first company approved to market a gene therapy medication. China's State Food and Drug Administration (SFDA) licensed Gendicine for treatment of head and neck squamous cell carcinoma (HNSCC).

In China, there are about 2.5 million new cancer patients every year. An estimated 7 million receive medical treatment. Company representatives project that 3% of the 7 million cancer patients would try Gendicine, as there is no other competitive drug available in the Chinese market at present (i.e., 210,000 potential customers).

Gendicine is sold at around 3,000 yuan (US$362) per injection. An average cancer patient needs to use six to eight injections, as SiBiono GeneTech's chief, Peng Zhaohui, explained. The annual market value for this particular drug amounts to US$532,140,000.

Meanwhile, the privileged status of Gendicine as the only gene therapy treatment available in China may not last long. H101 Adenovirus Injection, another biodrug against cancer developed by Shanghai Sunway Biotech Co. Ltd is in its third phase of clinical trials. Sunway's representatives expect that H101 will get SFDA drug authorization shortly.

The annual gene therapy market worldwide is expected to reach US$9.9 billion by 2007. In China, the annual market for the therapy was at least 4 billion yuan (US$483 million) at the beginning, Peng estimated. The biopharma's sales revenue was 6% of the total Chinese pharmaceutical industry revenue in 2000, and was over 24 billion RMB Yuan (*ca.* 2.4 billion Euro) in 2001.

The market is expanding and structuring itself, following the distribution of available wealth and technology, emerging brands, health insurance systems and new marketing strategies.

According to a report from the Chinese National Center for Biotechnology Development (Wang, 2003), 21 recombinant pharmaceuticals such as recombinant interferon, insulin and GCSF have been commercialized since China's first genetically engineered drug (recombinant human Interferon a1b) was brought to market in 1993 (Table 1).

As of 2002, in clinical trial stage, were more than 150 biopharmaceuticals (30 of which had type A New Drug status), 7 proprietary gene therapy drugs

Table 1. Biotech firms with gene therapy products for cancer in phase 2 or later of clinical development

Company or research institute	Indication	Delivered gene	Vector	Phase of clinical develop-ment
Shenzhen SiBiono Gene Technologies (Shenzhen, china)	HNSCC	Tumor protein p53	Adenovirus	Approved
Shanghai Sunway Biotech (Shanghai, China)	HNSCC	HAdv5 oncolytic virus	Adenovirus	Phase 3
AnGes MG (Osaka, Japan)	Arteriosclerosis obliterans	Hepatocyte growth factor	Plasmid	Phase 2
GenVec, Inc. (Gaithersburg, MD, USA)	Pancreatic, esophageal and rectal cancers	Human tumor necrosis factor	Adenovirus	Phase 2
Introgen (Austin, TX, USA)	Head and neck, lung, breast, esophageal, ovarian, bladder, brain, prostate and bronchoalveolar cancers	Tumor protein p53	Adenovirus	Phases 1–3
Transgene (Strasbourg, France)	Cervical cancer	Human papilloma virus type 16 E6 and E7 antigens and interleukin 2	Vaccinia virus	Phase 2
Transgene (Strasbourg, France)	Breast, lung, prostate and renal cancers	Human mucin 1antigen and interleukin 2	Vaccinia virus	Phase 2

including those for malignant tumor and hemophilia B, and 6 tissue engineering products (bone, cartilage, skin, tendon, etc).[1]

Sunshine Pharmaceutical Co., Ltd. (Chinese name: Shengyang Sanshen pharmaceutical Co. Ltd.) (*www.3sbio.com*) was founded in 1993. In its pipeline are products including therapeutic monoclonal antibody, recombinant peptides, DNA vaccine and molecular diagnostic products, according to the company.

Other gene therapy medicines in clinical trial in China include Thymidine Kinase Cyto by Shanghai Institute of Cancer, Recombinant adeno-associated Virus-2 Human Factor by Beijing-based AGTC Gene Technology Co. and Recombinant Interleukin-2 Adenovirus Antitumor Injection by Chengdu Centre of Gene Technologies.

For the time being, Peng Zhaohui argues that the upcoming competition is not a major threat to Gendicine. Technically, these medicines still need more time to get SFDA approval. SiBiono could utilize this period of time to develop more uses for Genedicine and new products related to the gene therapy medicine. In 2004 there were more than 700 gene therapy medicines in clinical trial.

2.3. Clinical trials

Advantages in conducting clinical trials in China include access to a very large number of patients or subjects, in addition to low cost. For example, in Southern China, nasopharyngeal cancer has an annual incidence of 10–150 (with an average of 80) per 100,000 population, compared to 5–9 (7) in Northern China, 15–20 (18) in Alaska and Greenland, and 1 in North America, Western Europe and Japan [3]. (Other sources refer to total rates of 2550 per 100,000 in China [4] From another perspective, of the 57,500 new cases that occurred globally in 1990 almost 45% was from China [5]. Whereas consumption behaviours and environmental factors have a causal impact on the distribution of this malignant tumour [3,63], genetic patterns have been described as significant [4,5].

Ideally, China's huge patient pool makes organization of trials faster and consequently shortens time in which the statistically meaningful data can be collected – hence shortening time to market. This was cited by SiBiono as one of the reasons for its being the first with a gene therapy drug on the market [7]. Surprisingly, however, observers noted that only 120 human subjects were enrolled in the final trial stage by Peng's team.

As far as Peng is concerned, the success story of his gene therapy enterprise has but begun (cf. "The Genesis of Gendicine: The Story Behind the

[1]These numbers looked slightly different according to the "Chinese Biotechnology Industry Development Report 2002". According to this report, about 30 biopharmaceutical drugs were in clinical trial phase up till 2002 and about 100 biopharmaceutical drugs and health care products were in R&D stage.

First Gene Therapy. BioPharm International exclusive interview," *http://www.biopharm-mag.com/biopharm/article/articleDetail.jsp? id=95485&pageID=1&sk=&date=*, site visited June 8, 2005).

He claims that, in May 2003, after 14 years, the results from phase 2 and 3 trials showed complete regression of tumours in 64% of the 135 patients with late-stage HNSCC who took part from November 2000 after eight weekly intratumoral injections of Gendicine in combination with radiation therapy 29% of the patients experienced partial regression. Among all the patients, about 75% suffered from advanced nasopharyngeal carcinoma, which is a sub-indication of head-and-neck cancer.

Accordingly, another 240-plus patients with late-stage HNSCC or terminal-stage non-HNSCC tumours were treated with Gendicine during the period from June 2003 to the present, continuing the phase 2 and 3 clinical trials. A patient receives one injection per week for four to eight weeks consecutively as a treatment cycle. A standard dose is 1×10^{12} viral particles (VP). Like the previous data, these trials also are said to show the safety and efficacy of Gendicine. Peng concludes that, in combination with chemo- and radiotherapy, that is proven effective to some degree especially in nasopharyngeal cancer treatment [8,9], Gendicine can improve treatment efficacy in symptomatic terms by a quantitative factor of 3.4. Furthermore, this combination appears to alleviate the toxic side effects normally associated with chemotherapy and radiation therapy.

Notably, all data narrated here from Pengs research still await scientific evaluation and further corroboration. As of today, no scientific publication of the experiments has appeared in a relevant international journal. Peng maintains that seven scientific papers reporting on the safety and efficacy of the clinical trials have been published in the National Medicine Journal of China (10 December 2003). This journal is not rated among international scientific standard literature. He also pledges international publication forthcoming any time soon.

Although Gendicine has been formally approved by SFDA only for indications of HNSCC, terminal patients with no other avenue of treatment have been allowed, on a case-by-case basis, to receive Gendicine (with permission from the SFDA). Required were a request from the patient, the patient's family, and the agreement of the patient's doctor. The ethics protocol includes involvement of the patient, family, and the doctor-in-charge in giving informed consent. The relevant ethics protocols have been drafted by Pengs group and accepted by the authorities, although they appear to contradict the principles of recent Chinese bioethical regulations, which generally accept the international standard model of individual's informed consent, rejecting the formal participation of relatives (see Chapter 3). Peng Zhaohui is optimistic that Gendicine will be approved for a wider range of cancer indications: Clinical trials seem to corroborate that Gendicine can be used to treat cancers of the digestive tract (esophageal, gastric, intestine, liver, pancreas, gallbladder, rectum), lung cancer, sarcoma, thyroid-gland cancer, breast cancer, cervical cancer, and ovarian cancer. The resulting market potential would be significant.

Peng's talk of success also builds upon the low incidence of relapse. According to his unpublished results, during more than three years of follow-up for the 12 patients with mid-to-late-stage laryngeal cancer who received Gendicine therapy in phase 1 clinical trials, no patient has relapsed. By contrast, among all patients who received only surgery, the three-year relapse rate is given by Peng as approximately 30% (with no more detailed data).

Although, statistically, 80% of the worldwide cases of nasopharyngeal carcinoma (a sub-indication of head-and-neck cancer), are in China, Peng claims that requests for Gendicine are also coming from the US, Germany, Denmark, Thailand, the Philippines, Greece, Canada, the UK, Singapore, Russia, Rumania, and Turkey. These patients are most likely of Chinese or Alaskan origin. Patients must come to China to be treated because the license is only valid in China (as to the profile of these patients, no data are available). In April 2004, about 400 patients had received the new treatment.

The development has been hailed in China as another sign of how the country is forging ahead in scientific research. However, the approval of Gendicine also provoked criticism because it is rather early to bring gene therapy to patients, especially the surprisingly small number of trial participants has raised concern about the scientific basis of the outcomes, although it should be noted that a small sample size does not necessarily indicate a flaw in the method. Still, a commentary noted that, "With 300,000 new cases of head and neck squamous cancers each year, China clearly thinks the benefits outweigh the risks" [10]. If Gendicine proves successful, SiBiono plans to launch the therapy in south-east Asia before seeking approval elsewhere.

2.4. Regulations and success

So why did the first commercial gene therapy treatment get produced and approved in China? "In China, where hundreds of thousands die of diseases such as cancer without access to the clinical options available to patients in the US and Europe, the potential for a one-time treatment that is relatively simple to administer is very appealing," explains Mark Kay, a director of the human gene therapy program at Stanford University (USA).

Public awareness of potential problems in gene therapy is just emerging. To my knowledge, there is no relevant scientific literature about the risk perception of Chinese people in the area of gene therapy. China has not been confronted with fatal failures, such as happened in the United States with the death of Jesse Gelsinger, who died from side effects of a gene therapy trial and more recently in Europe with the X-linked, severe combined immunodeficiency syndrome trials. Hence, it is assumed that the Chinese regulatory authorities may be more receptive to the potential benefits of the technology and less alert towards the risks, until positive evidence suggests revision of policy.

Foreign competitors and commentaries have suggested that the regulatory process is less strict in China than elsewhere. "The recently approved gene therapy in China had only 120 people in clinical trials, whereas the same therapy in the US has hundreds of people and yet it has not been approved," says Hitoshi Kotani, senior vice president of gene therapy firm AnGes MG (Osaka, Japan). In a recent article dedicated to China as a test site and future market for new drugs, *Nature* magazine even speaks of an "ethical mire", regarding patients and trial subjects' rights, and refers to "the wilds of Chinese clinical research" [11]. Chinese bioethicists have rejected this statement as biased and poorly informed.

Researcher and entrepreneur Peng refutes that Gendicine was approved because of the allegedly looser regulation of the Chinese authorities. This sentiment is echoed, e.g., by Peng Shang, vice director of the Cell Engineering Research Center at the Fourth Military Medical University (Xi'an, China). "In fact, the SFDA had a routine practice not to approve any new kind of medicine if the kind of drug was not authorized by the US FDA," Shang says. Peng has been lobbying SFDA for years and the agency gradually changed its attitude, as shown by the approval of the new gene therapy and, for example, by giving a green light to clinical trails for a new SARS vaccine developed by Sinovac (Beijing) [12].

In general, expectations of combined economic and health-related benefits, with a special notion of international competition, and the obvious need for greater flexibility and efficiency as acknowledged through the SARS crisis, might have helped policy makers to adopt a less conservative attitude. Pressures on the administration, not to stand in the way of urgently needed good news in medicine and the health sector, provides keen researchers with new opportunities.

3. ON CULTURAL EXPECTATIONS

Interpreting legislation in the area of gene therapy can benefit from related research in China's bioethics. Let us now consider the debate about bio-policy making and its "cultural" dimension, in the area of human embryo stem cell research.

3.1. Culture misconceived

The assumed connection between cultural and research factors has been stated in a straightforward manner. For example, an alleged cultural peculiarity is propagated as a competitive advantage, in a special issue of *Nature*.

"Therapeutic cloning, stem-cell studies and other research areas that use animal or human embryos are controversial and raise religious and ethical questions (...).

These issues have led to unsupportive policies for cloning-related research, and the high costs of clinical trials for any proteins developed using this technology have forced many scientists and commercial companies to abandon promising research and to lose out on potentially profitable products.

China has a cultural environment with fewer moral objections to the use of embryonic stem cells than many Western countries, and (...) it could take a leading role in this field (...).

China has probably the most liberal environment for embryo research in the world (...).

In addition, the relatively easy access to human material, including embryonic and fetal tissues, in China is a huge advantage for researchers."

Together with China's cultural characteristics, "these technologies offer unprecedented research and commercialization opportunities for China" (Yang, 2004).

It is obvious that reference to culture can be more or less accurate. In these cases it plainly disguises vital stakes, such as in the competition between researchers for fame and funding. More often than not, culture is used as a magic stick to shun rational analysis and reason-guided discourse, which would uncover prevailing profane interest. Here, "culture" becomes an ideological pattern in the fabric of the frame to protect the purported humanitarian mission of biomedical and biotech research.

This observation indicates that "cultural arguments" require an educated assessment. It alerts us about the issue of sincerity among the proponents of "cultural diversity" and the "purity" of science.

3.2. China's approximation of ethical demarcations

However, proper research on this area reveals that, in sum, Chinese authorities establish a regulatory and ethical framework in relation to human embryonic and fetal stem cells that reflects emerging international standards. In the bioethical field, China is working towards a recognizable liberal European framework of regulation, in several cases based on the British House of Lords Select Committee recommendations. The main concern is, that it remains unclear, how fully such standards are accepted and reflected in practice outside the centres of international excellence.

According to findings from my own research, China's biopolitics can be characterized by two general moral demarcations, with notable Chinese peculiarities beyond pragmatic considerations of global harmonization of standards.

(1) *A Chinese Rubicon* defines the beginning of human worthiness of protection or dignity. The local transfer of an embryo, from the petri dish

into the uterus, demarcates the line between research and medical or invasive treatment. Manipulation *in vitro* might be permitted, but implantation into the female system is a taboo. (2) *A Chinese Limes* in bioethics is defined in terms of the social–moral dimensions that constitute the human being. As soon as a human is born alive, into the social environment, the full power of legal protection begins to apply.

3.2.1. A Chinese Rubicon

The Chinese Rubicon is based on a strong notion of natural purity and dignity, which can be traced back in the history of Chinese philosophy to Neo-Confucianism. Reference to other philosophical sources of naturalism are made, e.g., to the Han-dynastic amalgamation of cosmological, social–moral and political concepts, especially a sexualised interpretation of Yin and Yang [13,14].

According to an elaborated contribution on a New-Confucian background, biotechnology is, in principle a legitimate human endeavor. According to Taiwanese philosopher, Lee Shui-chuen, "assisting nature", e.g., biomedically, includes "purification" of humanity *and* the world. Moral quality and natural constitution are interconnected [25,26]. The natural constitution of an embryo may be altered if certain conditions apply, but, according to the Beijing school around bioethicist Qiu Renzong, an altered embryo may not become part of the causal chain, or, the human social–biological system. Products of hybridization and other forms of manipulation must remain inside the dish.

This *Imperative of Purity* already gained regulatory force in reproductive medicine (prohibition of nuclear transfer or ooplasma transplantation). The use of cloning technology for human reproduction is unlikely to be endorsed in China.

3.2.2. A Chinese limes

The second line of moral demarcation has a legal form. It can be adjusted through political or social process. This Limes in bioethics is defined in terms of the social–moral dimensions that constitute the human being. When a human is born alive legal protection begins to apply. It is the onset of a gradual development of the social career, in the course of which nobody may be manipulated or killed (exceptions apply).

Without the psycho-emotional and physical relationship with a mother, early in development, preceding immediate social relations, an embryo, accordingly, is "only a human life but not a full social entity". It may be taken as a commodity for "high ranking medical purposes".

Generally, any action in medical context is regarded as a challenge, because it interferes with the course of nature and humanity. Thus, it assumes a practice that can, according to Confucianism, only be legitimate when interfering takes place in terms of "assisting Heaven and Earth", that is, not altering nature's course.

From a cultural perspective, both demarcations cannot be played against each other, nor can one be neglected. They constitute major elements in the fabric of cultural context.

Thus, the issues of moral status of human beings are not solved. In effect, the "embryo matter" is brought to the level of individual social relation clusters and taken out of the charge of the public and experts. It leaves room for a plurality of moral practices, without inviting ethical relativism.

The moral culture in China, as much as it can be described and as far as it is relevant for these issues, is far from being fundamentally at odds with moral views in European and North American countries. The moral landscape between the Chinese Limes and Rubicon is diverse.

Considering the adaptability of international bioethical regulations in China, consequentially, no major obstacles are expected from China's politics and administration, as far as technical issues and standards are concerned. The relevant Chinese policies respond to the pragmatic demand for harmonization of international standards. Practical implementation and monitoring of biopolicy, however, raises concern. Full assurance of respect of the powerful towards each individual's basic rights, born or unborn, is still a dream of the future.

Having said this, a word of caution should be in place. Even when described in fair and accurate manner, "culture" does not always help us to understand the factors involved and the major guiding motives in the expression of ethical norms. Nor is it always appropriate to discuss developments within the specific focus of an explicit cultural frame. Especially in areas of global competition and fast growth of vital stakes, as it happens in the case of biomedicine in China, the most articulate normative contributions are programmed to manifest pragmatic policy making in the light of the harmonization of international standards.

The theme of culture, though, is always present. But the proper level of attention to and emphasis on "culture" needs to be established.

4. THE ROLE OF RESEARCHERS IN ETHICS REGULATIONS AND THE INTERNATIONAL STATE OF THE ART: CHINA'S POSITION

It is part of the strategies of biotechnology-related policies to ally with internationally acclaimed researchers. James Watson, Francis Collins and other celebrities, for instance, have expressed admiration and support in particular for the genomics' activities organized by the Huada Genome Center at Beijing, under the directorate of Yang Huanming, with branches in Hangzhou and Xi'an. This ensured political backing in times of domestic competition and political uncertainty (e.g., in 1998, the government was undecided, as to whether to adopt the HGO "public ownership" policy or Craig Venter's commercial approach to the sequencing of the Human Genome. These camps in China were represented by Yang Huanming and Chen Zhu, respectively).

Notably, Yang Huanming served as China's delegate to the UNESCO's IBC that drafted the relevant document for universal ethical standards in bioethics, such as the "Universal Declaration on the Human Genome and Human Rights", that is, the only international instrument in the field of bioethics, which was endorsed by the United Nations General Assembly (in 1998). Yang has been perhaps the most explicit and influential Chinese life scientist in the area of bioethics. In particular his engagement in the domestic movement of critics against the infamous "Mothers and Infants Health Care Law" (Eugenics Law, of 1995) (*Yousheng*) [15,16] has earned him international and domestic acclaim, supporting his humanistic call for responsible science.

In the case of Gendicine, gene therapy icon French Anderson, agreed to function as an unpaid adviser to SiBiono. His credentials confirmed the company's and researches' international standing. Anderson is quoted with a statement that is both, scientifically supportive and an exercise in a foreigner's politeness and a grain of symbolism that certainly was appreciated by his partners. Noting that the company's adenovirus is relatively simple compared with the gene-delivery systems being developed in the west. He offered, "but sometimes simple is better" [10].

In the absence of an explicit legal framework for his trials, Peng resorted to a method that had proven successful in other areas of the life sciences in China. He initiated the making of the relevant regulations, which would then be approved of and used as standards by the authorities in charge.

Here is an excerpt from the regulatory standards as they were accepted for the purpose of licensing for marketization of Gendicine (Taken from: http://www.biopharm-mag.com/biopharm/article/articleDetail.jsp? id=95486& pageID=1&sk=&date=; (visited June 8, 2005).

"Points to Consider for Human Gene Therapy and Product Quality Control State Food and Drug Administration of China. This document is authored by Shenzhen SiBiono GeneTech Co., Ltd. and the National Institute for the Control of Pharmaceutical and Biological Products, China. It is now SFDA's official guidance regarding gene therapy development. It was translated from the Chinese by Shenzhen SiBiono GeneTech Co., Ltd. May 1, 2004 by: Zhaohui Peng, State Food and Drug Administration of China, BioPharm International"... 8. Ethics study.

Special attention should be paid to medical ethics during clinical trials of gene therapy products. Details can be found in SFDA's GCP (Good Clinical Practice) regulations.[2] The study plan and potential risks associated with the

[2] SFDA guidelines for good clinical practice. Beijing, 1999, Sep 1.

clinical study should be clearly communicated to the patient and family members. Patients have the right to choose medical treatment options and terminate participation in the gene therapy clinical trial. The patient's medical history should be kept private. The patient cannot be enrolled in the study until the patient or family member signs the study consent form.

(These formulations are obviously sloppy and fall short of the standards set by other Chinese regulations, especially the "informed consent" regime that has been reformed according to international standards, namely stipulating that individual consent must be taken on the basis of strict non-coercion or compensation, full information of the patient/trial participant, anonymity of personal data and privacy protection.)

In the area of human embryo research and the related clinical activities, two outstanding females have contributed significantly to the enhancement of legislation. Lu Guangxiu and Sheng Huizhen serve as examples for individual leading scientists taking the initiative in making guidelines in the form of proposals, which would eventually be accepted as binding on the level of policy making.

Lu Guangxiu is a fertility doctor and medical researcher with a long career in both medicine and the political establishment of the central Chinese province of Hunan. It is generally known that she has taken the leading role in advising the process of legislation related to fertility, clinical research, IVF, embryo storage, and egg donation and sperm banking. Her internal rules stood model for the current state of national legislation. For example, the so-called "Human-Yale Pattern" [17] describes a practice of patient–physician co-operation that is designed to account for the peculiarities of Chinese families and culturally based attitudes towards reproduction and family matters. The current regulations of reproductive medicine and sperm banks largely result from Dr. Lu's experience as a second-generation reproductive doctor and her interests as well as concerns as a physician, researcher and policy maker [18].

When embryologist Sheng Huizhen returned to Shanghai from her research in the USA she brought alongside with her scientific excellence profound insight into the vital role of internationally accepted basic ethical regimes. In Shanghai, she found a situation that was characterized by an absence and an opportunity at the same time. The National Genome Center at Shanghai had established an ELSI group that was administratively connected with her Xinhua clinic (belonging to Shanghai Second Medical University). At that time, in 2001, there were no regulations on embryo research in place, except for a general and vague announcement by the Ministry of Health, rejecting human cloning, cross-species hybridization and "research contradicting common morality".

Sheng's personal intervention and research agenda eventually inspired the ELSI group to propose a set of guidelines, which were the first politically endorsed detailed standards for research on human embryos in China [19].

5. WHAT IT MEANS FOR US?

This overview has hopefully illustrated that a general reference to "culture" is misleading and not informative for the purpose of understanding bioethics in China properly. More empirical evidence must be provided, relevant levels of legislation and discourse must be distinguished and their content be protected from premature or out of context references to "culture".

Focused scrutiny on this area of society that is involved with policy making and the formulation of positive standards provides a realistic and detailed view of important patterns in the fabric of the country's normative culture, but it is still not adequately representing China's culture in her depth and wealth. It should be emphasized here, that in China, in particular, bioethics is not organized as a democratic or otherwise transparent and representative process, not to mention a discourse. This structural characteristic indicates the most imminent problem for bioethicists to face in China, from the perspective of ethics and in view of social sustainability of the life sciences [20]. Lack of transparency is detrimental to the overall political and ethical goal of trust, because it contradicts efforts to enhance individual and institutional accountability and responsibility. Evidently, this lack of transparency can be found in three interrelated areas. The scientific and medical facts are not clear and hence impossible to assess properly; the knowledge, understanding, communication and practice of ethical standards in this important area of the life sciences is inadequate (it is difficult to tell to what an extent this is owing to lack of compliance or insufficient knowledge); the practice of approval in this particular case indicates that authorities are challenged by opportunism and probably not fully trained and thus fail to implement state-of-the-art legislation. Altogether, the relevant process seems to take place inside a "black box" of decision-making, which creates grey areas of practice, inviting speculations and creating a discomforting public a scientific image.

European life scientists and ethicists have hardly begun to pay attention to China in this regard. In order to establish a culturally informed and sensitive bioethics research, more co-operative studies with Chinese colleagues are needed. They should obviously adhere to the fundamental scientific virtues of descriptive accuracy and methodological soundness. They should account for the real complexities of a heterogeneous society in transformation, that is, they should expect prescriptive uncertainty and be prepared to support ethics and legislation, even in places where they are not fully adhered to. This requires resolve in the ethics of science, especially self-discipline when facing the allure of non-scientific interests and to avoiding theoretical bias, such as in terms of culturalism.

Culturalistic flaws can be avoided if we methodically highlight the significant role of individual actors and institutions within the process of legislation. We can also benefit from systematic practical interaction, in the spirit of critical partnership, with Chinese colleagues; given the intricacies of European bioethics this is certainly not a one-way street.

It has been the purpose of this paper to show an approach that can help to avoid cultural biases and presuppositions in a spirit of cultivated and interdisciplinary science, namely by distinguishing interpretative contexts, actors, and account for their respective relevance within a system that is as much in flux as it deserves serious attention.

REFERENCES

[1] R. P. Suttmeier and X. K. Yao, China's Post-WTO Technology Policy: Standards, Software, and the Changing Nature of Techno-nationalism, NBR Special Report, The National Bureau of Asian Research, 7, 2004.

[2] Y. Tang, Biotechnology in China. A Guide to the Chinese Biotechnology Industry, Dechema Report, 2004.

[3] H. Cheng, Nasopharyngeal cancer and the southeast Asian patient American family physician, 2001, **63**(9), 1776–1782.

[4] D. B. Goldsmith, T. M. West and R. Morton, HLA associations with nasopharyngeal carcinoma in Southern Chinese: A meta-analysis, *Clin. Otolaryngol. All Sci.*, 2002, **27**(1–61).

[5] P. N. Notani, Global variation in cancer incidence and mortality, *Curr. Sci.*, 2001, **81**(5), 465–474.

[6] R. W. Armstrong, M. J. Armstrong and M. S. Lye, Social Impact of nasopharyngeal carcinoma on Chinese households in Selangor, *Malaysia, Singapore Med. J.*, 2000, **41**(12), 582–587.

[7] S. Pearson, H. P. Jia and K. Kandachi, China approves first gene therapy, *Nature*, 2004, **22**(1), 3–4.

[8] S. H. Cheng, J. J. Jian, S. Y. Tsai, K. Y. Chan, L. K. Yen and N. M. Chu, Prognostic features and treatment outcome in locoregionally advanced nasopharyngeal carcinoma following concurrent chemotherapy and radiotherapy, *Int. J. Radiat. Oncol. Biol. Phys.*, 1998, **41**, 755–762.

[9] S. H. Cheng, S. Y. Tsai , K. L. Yen, J. J. Jian, N. M. Chu , K. Y. Chan *et al.*, Concomitant radiotherapy and chemotherapy for early-stage nasopharyngeal carcinoma, *J. Clin. Oncol.*, 2000, **18**, 2040–2045.

[10] S. P. Westphal, First gene therapy approved, *New Scientist*, 2005. 28 November 03 (www.eurekalert.org/pub_releases/2003-11/ns-fgt112603.php or www.newscientist.com) [Accessed 12 May 2005]).

[11] D. Cyranosky, Consenting adults? Not necessarily ..., *Nature*, 2005, **435**, 138–139.

[12] Z. H. Wang, SARS vaccine approved for human trials, China Daily, Aug. 6, 2005 (http://www2.chinadaily.com.cn/english/doc/2005-08/06/content_466753.htm [Accessed 19 August 2005]).

[13] R. Z. Qiu, Cloning issues in China, in *Cross-Cultural Issues in Bioethics: The Example of Human Cloning* (ed. Heiner Roetz), Amsterdam/New York (Rodopi), 2005.

[14] O. Döring, Culture and bioethics in the debate on the ethics of human cloning in China, in *Cross-Cultural Issues in Bioethics: The Example of Human Cloning* (ed. Heiner Roetz), Amsterdam/New York (Rodopi), 2005.

[15] O. Döring, Eugenik' und Verantwortung: Hintergründe und Auswirkungen des Gesetzes über die Gesundheitsfürsorge für Mütter und Kinder, *China aktuell*, 1998, **8**, 826–835.

[16] D. Dickson, Congress grabs eugenics common ground, *Nature*, 1998, **394**, 711.

[17] L. J. Li and G. X. Lu, How medical ethical principles are applied in treatment with artificial insemination by donor in Hunan, China: Effective practice at the reproductive and genetic hospital of CITIC-Xiangya, *JME*, 2005, 333–337.
[18] O. Döring, China's struggle for practical regulations in medical ethics, *Nat. Rev. Genet.*, 2003, **4**, 233–239.
[19] O. Döring, Chinese researchers promote biomedical regulations: What are the motives of the biopolitical dawn in China and where are they heading? *Kennedy Inst. Ethic J.*, 2004, **4**(1), 39–46.
[20] O. Döring, *Chinas Bioethik verstehen. Ergebnisse, Analysen und Überlegungen aus einem Forschungsprojekt zur kulturell aufgeklärten Bioethik,* Abera Verlag Hamburg, ISBN, 3-934376-58-4, 2004.
[21] O. Döring, Der menschliche Embryo in China: im Spannungsfeld zwischen Forschungsmaterial, Fürsorge und Charakterfrage", in Fuat S. Oduncu, Katrin Platzer, Wolfram Henn (Hrsg.): Der Zugriff auf den Embryo. Ethische, rechtliche und kulturvergleichende Aspekte der Reproduktionsmedizin. Reihe Medizin-Ethik-Recht Band 5 (hg. v. Oduncu FS, Schroth U, Vossenkuhl W), Göttingen (Vandenhoeck & Ruprecht), 2005, pp. 126–145.
[22] O. Döring, Bioethics and sustainable life sciences. A programmatic reflection, in *Proceedings of the International Congress of Bioethics* (ed. H. Mohammad), (Tehran March 26–28, 2005), Bitrafang Publishing, 2005c, pp. 259–274.
[23] H. P. Jia, First gene-therapy medicine commercialized. China business weekly [online]. Beijing, China. China Daily, 2003. Available from: http://www/chinadaily.com.cn/en/doc/2003-12/09/content_289867.htm [Accessed 19 January 2004].
[24] H. P Jia, Gene therapy finds welcoming environment in China. Are China's regulations too lax? *Nat. Med.* (Published online: 1 March 2006; | doi:10.1038/nm0306-263).
[25] S. C. Lee, A Confucian Perspective on Human Genetics, in *Chinese Scientists and Responsibility*, (ed. Ole Döring), Mitteilungen des Instituts für Asienkunde Nr. 314, Hamburg, 1999, pp. 187–198.
[26] S. C. Lee, A Confucian Assessment of 'Personhood', in *Advances in Chinese Medical Ethics. Chinese and International Perspectives* (ed. Ole Döring and Chen Renbiao), Hamburg, 2002, 167–177.

Does Patent Granting Hinder the Development of Gene Therapy Products?

Clara Sattler de Sousa e Brito

[1] *Max Planck Institute for Intellectual Property, Competition and Tax Law, Germany*

Abstract

Each year there are approximately 55,000 cases of breast cancer diagnosed in Germany, and approximately 18,000 women die from the disease. About 5–10% of breast cancer cases are caused by hereditary genetic alterations in a few certain genes, mainly the so-called breast cancer genes BRCA1 and BRCA2 (**br**east **ca**ncer **1** and **br**east **ca**ncer **2**) discovered in the early 1990s. For women with an inherited alteration in one of these BRCA genes, the risk of developing breast cancer during their lifetime increases from 10% to 80%. In other words, women with an altered BRCA1 or BRCA2 gene are up to eight times more likely to develop breast cancer than women without alterations in those genes. The US Company Myriad Genetics, Inc. owns 260 patents worldwide on BRCA1 and BRCA2, either on the genes themselves or diagnostic methods based on them. The European Patent Office (EPO) granted four patents to the US company from Salt Lake City/Utah on these genes (EP 0 699 754, EP 0 705 902, EP 0 705 903 and EP 0 785 216). The patenting of the breast cancer genes BRCA1 and BRCA2 and of the predictive molecular genetic diagnostic test based on the genes is highly controversial. Worldwide concerns have been raised with respect to the quality of the test, the access for patients to the test and a hindrance to research and development. In particular, it is expected that research and development activities for promising therapies such as gene therapies based on these genes and the production of new drugs may be hindered. This paper will start by showing the importance and significance of the problems associated with patenting disease-associated genes in general. Then, by means of the example of the BRCA genes, it will discuss to what extent the concerns are well founded and which possibilities exist to thwart negative implications related to the patenting of disease-associated genes.

Keywords: patent, patent and genetherapy, BRCA1 and BRCA2, patents on BRCA1 and BRCA2, disease associated genes, genetic testing, priority, novelty, inventive step.

Contents

1. Public discussion 160
2. Basic facts about patents 160
3. Patent law and gene therapy 161
    3.1. Patents granted on disease-associated genes 162
        3.1.1. Colorectal cancer 162
        3.1.2. Hemochromatosis 163

3.1.3. Canavan disease 163
3.1.4. Breast cancer 164
3.2. Patents belonging to myriad genetics 165
3.2.1. EP 0 699 754 – "method for diagnosing an predisposition for breast and ovarian cancer" 165
3.2.2. EP 0 705 903 – "mutations in the 17q-linked breast and ovarian cancer susceptibility gene" 166
3.2.3. EP 0 705 902 – "17q-linked breast and ovarian cancer susceptibility gene" 166
3.2.4. EP 0 785 216 – "chromosome 13-linked breast cancer susceptibility gene BRCA2" 166
4. The concerns 166
4.1. Costs of tests 167
4.2. Quality loss of testing 167
4.3. Quality loss of medical practice 168
4.4. Commercialisation of medical genetic testing 168
4.5. Further research growth and development of diagnostic methods and new therapies 169
4.6. Effects of licensing 170
4.7. Summary 170
5. Patenting conditions and opposition procedures in the case of the BRCA genes 171
5.1. Patent conditions 171
5.2. Opposition procedure 171
5.2.1. Lack of priority and absence of novelty 171
5.2.2. Lack of inventive step 172
5.2.3. Insufficient description 172
5.3. Decisions of the opposition division 172
5.3.1. EP 0 699 754 173
5.3.2. EP 0 705 903 173
5.3.3. EP 0 705 902 173
5.3.4. EP 0 785 216 174
6. Conclusion and outlook 174
References and notes 176

1. PUBLIC DISCUSSION

Although Patent Law has not been a matter of public concern for decades, at present it is part of a heated public discussion and for the first time, ethical concerns appear with radical importance in the debate. The advent of striking new possibilities in human medicine, such as genetic testing, gene therapy or reproduction medicine are accompanied with horror scenarios like the cloning of humans. This feeds the highly and already emotional discussion.

2. BASIC FACTS ABOUT PATENTS

According to the European Patent Convention (EPC), a patent is an exclusive right granted for an invention, which is a product or a process that provides a new way of doing something, or offers a new technical solution to a problem.

A patent provides protection for the invention for the owner of the patent. Patent protection means that the patent holder has, temporarily, exclusive rights to the exploitation of the invention in a given territory, that is, the patent holder can prohibit unauthorized third party commercial use, which means the invention cannot be commercially made, used, distributed or sold without the patent owner's consent. The protection is granted for a limited period, generally 20 years. The patent holder thus holds a temporary monopoly right to the exploitation of the patented invention, but once a patent expires, the protection ends and the invention enters the public domain. That is to say the owner no longer holds exclusive rights to the invention and it then becomes available to commercial exploitation by others.

Patents provide incentives to individuals by offering them recognition for their creativity and material reward for their marketable inventions. These incentives encourage innovation.

For biotechnological inventions the possibility of amortisation of costs is especially important, as research in this area is always highly expensive. The high capital investments only pay off if the company can protect their results by exclusive rights, especially patents, and thus keep their competitive advantage.

The important direct relation between patent protection and the economic wealth of the biotechnology industry was illustrated, for example, by a declaration in 2000 by Tony Blair and the former US president, Bill Clinton, which states that scientists around the world should have free access to research on the mapping of human genes [1]. This was understood to be a way of preventing the patenting of human genes and following this political move, the stock drive of Celera Genomics and other biotechnology companies lost up to 20% in only a few minutes.

Patents, however, provide more than incentives to individuals by ensuring recognition and material reward. They also help to enrich the total body of technical knowledge in the world. Patent owners are obliged to publicly disclose information on their invention; thus providing valuable information for other inventors, as well as inspiration for future generations of researchers and inventors. At the same time, the publication opens the possibility for the society to know what research is actually going on, and furthermore which direction it is going to take in the R&D labs thereby allowing it to react both by means of law as well as economic and scientific policy. Thus, it appears that Patent Law is an extremely powerful tool capable of directing science.

3. PATENT LAW AND GENE THERAPY

The history of gene therapy clearly shows the close relation between therapy hopes, scientific progress and patenting [2]. At the beginning, vast amounts of money were invested by large pharmaceutical companies in collaborations with, or even through acquisitions of small biotech companies owning patents in the area of gene therapy.

An especially intriguing example is the takeover for $295 million of the company Genetic Therapy Inc. Gaithersburg, MD, USA by the Swiss pharmaceutical company Sandoz, AG, Basel in 1995. Genetic Therapy Inc. was essentially taken over for its ownership of an exclusive license for a (already at that time extremely controversial [3]) US patent [4] covering the whole range of somatic gene therapy.

Dr. Daniel Vasella, at that time the president of the Sandoz pharmaceutical section, now Chief Executive Officer (CEO) of Novartis commented on the deal with the following statement: "We strongly believe gene therapy will transform 21st century medical practice. We concluded that owning this technology would be very important to Sandoz" [5].

Despite the $295 million takeover and the initial enthusiasm, about eight years later Novartis closed down Genetic Therapy as the expected results failed to materialise [6].

This shows, first, how costly investments in new therapy methods are, and second, how important patents are to encourage investments like this and lastly how risky these investments in new technologies can be [7].

3.1. Patents granted on disease-associated genes

In the case of the patenting of disease-associated genes, there are mainly concerns that patenting might constrain research and development. These concerns were discussed worldwide in relation to the patenting of the genes BRCA1 and BRCA2 (**br**east **ca**ncer **1** and **br**east **ca**ncer **2**), the so-called breast cancer genes. The concerns regarding the patenting of the BRCA genes gain more weight when one considers the fact that, besides the patents on BRCA1 and BRCA2, other patents on disease-associated genes were granted from the EPO. Some of these relevant disease-associated genes are ones for hereditary colorectal cancer, hemochromatosis or Canavan disease (CD). After a short description of the situation in these diseases, as the most well known and heavily debated example, the case of breast cancer will be analysed more closely.

3.1.1. Colorectal cancer

There are approximately 66,000 cases of colorectal cancer diagnosed in Germany each year, and approximately 29,000 deaths in men and women. Colorectal cancer is the second most common cancer and the second leading cause of cancer deaths. An average of 5 out of 100 people in Germany will develop colon cancer in their lives. About 5–10% cases of colorectal cancer are caused by genetic alterations in a certain gene. There are two main distinguishable clinical forms of the disease: hereditary non-polyposis colon cancer (HNPCC) and familial adenomatous polyposis (FAP).

The most common type of inherited colorectal cancer is HNPCC. It accounts for about 5% of all colorectal cancer cases. It is caused by changes in DNA mismatch repair genes on chromosome 2, 3, 5 and 7. About 90% of the HNPPC cases are caused by mutations in the genes MLH1 and MLH2. About 80% of the people with an altered gene develop colon cancer over a lifetime, and the average age for diagnosis of HNPCC is 44, in contrast to 64 for the spontaneous form.

FAP is an inherited condition, which is caused by a mutation in a specific gene called APC located on chromosome 5 [8]. The lifetime risk of developing FAP is 90% with an altered APC gene.

US patents exist for the MLH1, MLH2 and APC genes, mutations of the APC gene and methods for the detection of these mutations. They are held by the Faber Cancer Institute, the John Hopkins University and the University of Utah, which licensed non-exclusively to Myriad Genetics.

3.1.2. *Hemochromatosis*

Hemochromatosis is a genetic disorder of the metabolism. Individuals with hemochromatosis cannot excrete iron; therefore, the excess builds to toxic levels in tissues of major organs such as the liver, heart, pituitary, thyroid, pancreas, lungs and synovium. Undiagnosed and untreated hereditary hemochromatosis (HHC) can develop into various diseases such as diabetes, heart trouble, arthritis, liver disease, neurological problems, depression, impotence, infertility and cancer.

In 80–85% of the cases, hemochromatosis is caused by a mutated allele of HLA-H-gene (C282Y and H63D). This makes genetic testing a very useful diagnostic tool, especially as hemochromatosis, due to the enormous broad range of symptoms, is difficult to diagnose with conventional methods.

The US patents which cover the HLA-H-gene itself [9] as well as the molecular diagnosis of HLA-H-gene [10] were granted to Mercator Genetics, Inc., who licensed this technology exclusively to SmithKline Beecham Clinical Laboratory (SBCL) for a payment of approximately $3 million.

In Europe, Mercator Genetics, Inc. also applied for two patents: one of them covering the HLA-H-gene itself [11] and the other one the diagnostic method based on the HLA-H-gene [12].

3.1.3. *Canavan disease*

Canavan disease (CD) is an inherited, degenerative brain disorder that, if both parents are carriers of the defective gene, affects children at a rate of one in four pregnancies (25%). It leads first to a loss of body control and finally to death usually before the children reach the age of 10. CD is caused by mutations in the gene for an enzyme called aspartoacylase, which leads to a deficiency of that enzyme. In 1997, the Miami Children's Hospital

(MCH) succeeded in cloning the Canavan gene and patents for the gene as well as protein and diagnostic screening methods were granted [13].

3.1.4. Breast cancer

Each year more than 55,000 German women learn that they have breast cancer and 18,000 die of breast cancer annually. Breast cancer is the leading cause of cancer deaths among women in Germany, and in all western developed countries [14].

Furthermore, the incidence of breast cancer appears to be inexorably increasing [15] with an annual worldwide incidence of over 1 million predicted at the beginning of the 21st century [16].

Hereditary breast cancer accounts for approximately 5–10% of all breast cancer cases and a large percentage of them are early onset breast cancer [17]. Mutations in BRCA1 and BRCA2 make some women more susceptible to develop breast and ovarian cancer.

Women with an inherited alteration in one of these BRCA genes have a greatly increased lifetime risk of developing breast cancer compared to the general population and a higher chance of getting it at a young age. Men with an altered BRCA1 or BRCA2 gene also have a significantly increased risk of breast cancer.

According to estimates of lifetime risk [18], about 10% of women in the general population will develop breast cancer, compared with estimates of up to 80% of women with an altered BRCA1 or BRCA2 gene. In other words, women with an altered BRCA1 or BRCA2 gene are up to eight times more likely to develop breast cancer than women without alterations in those genes.

Decoding of BRCA1. In 1990, the Group of Mary-Claire King localised the first breast and ovarian cancer predisposing gene on chromosome 17 [19]. This scientific breakthrough led to a race for the sequencing of the gene which in October 1994 was finally won not by King's group, but by the team of Mark Skolnick consisting of scientists from the University of Utah, the company Myriad Genetics, supported by the U.S. Institute of Health (NHI), McGill University and the company Eli Lilly [20].

Decoding of BRCA2. During the research on BRCA1, the existence of a second gene related to breast and ovarian cancer on chromosome 13 was found. Against all expectation, it was not Myriad Genetics but a British team under Richard Wooster and Michael Stratton from the British Institute for Cancer Research (ICR) in collaboration with the Duke University School of Medicine and Sanger Centre in Cambridge [21] who discovered it. After the first breakthrough in 1994, BRCA2 was localised in summer 1994 [22] followed by another important hint from scientists of John Hopkins University School of Medicine [23], finally leading to the sequencing of this second gene.

Who succeeded first in sequencing BRCA2 is highly controversial as two groups claimed the sequencing independently. The first one was the group of scientists from ICR under Stratton [24], the second group consisted of scientists from the University of Utah and Myriad Genetics.

3.2. Patents belonging to Myriad Genetics

The contents of the European Patents are well described by the following lines out of the US patent application:

> The present invention relates generally to the field of human genetics. Specifically, the present invention relates to methods and materials used to isolate and detect a human breast and ovarian cancer predisposing gene (BRCA1), some mutant alleles of which cause susceptibility to cancer, in particular breast and ovarian cancer. More specifically, the invention relates to germ line mutations in the BRCA1 gene and their use in the diagnosis of predisposition to breast and ovarian cancer. The present invention further relates to somatic mutations in the BRCA1 gene in human breast and ovarian cancer and their use in the diagnosis and prognosis of human breast and ovarian cancer. Additionally, the invention relates to somatic mutations in the BRCA1 gene in other human cancers and their use in the diagnosis and prognosis of human breast and ovarian cancer. The invention also relates to the therapy of human cancers which have a mutation in the BRCA1 gene, including gene therapy, protein replacement therapy and protein mimetics. The invention further relates to the screening of drugs for cancer therapy. Finally, the invention relates to the screening of the BRCA1 gene for mutations, which are useful for diagnosing the predisposition to breast and ovarian cancer. [25]

Myriad Genetics holds the following four European patents:

EP 0 699 754 – "Method for Diagnosing a Predisposition for Breast and Ovarian Cancer"

EP 0 705 903 – "Mutations in the 17q-Linked Breast and Ovarian Cancer Susceptibility Gene"

EP 0 705 902 – "17q-Linked Breast and Ovarian Cancer Susceptibility Gene"

EP 0 785 216 – "Chromosome 13-Linked Breast Cancer Susceptibility Gene BRCA2"

3.2.1. EP 0 699 754 – "Method for Diagnosing a Predisposition for Breast and Ovarian Cancer"

This patent granted in early 2001 to Myriad Genetics Inc., the University of Utah Research Foundation and the United States of America mainly relates to a method for diagnosing a predisposition for breast and ovarian cancer.

3.2.2. EP 0 705 903 – "Mutations in the 17q-Linked Breast and Ovarian Cancer Susceptibility Gene"

This patent was granted in May 2001 to Myriad Genetics, the Centre de Recherche du Chul, Canada and the Cancer Institute, Tokyo, Japan and contains some isolated nucleic acid and special mutations of it [26], related vectors [27], the production of the BRCA1 polypeptides [28] and its use for antibody production [29] as well as a method for diagnosing a predisposition for breast and ovarian cancer [30].

3.2.3. EP 0 705 902 – "17q-Linked Breast and Ovarian Cancer Susceptibility Gene"

This patent was granted in November 2001 to Myriad Genetics Inc., the University of Utah Research Foundation and the United States of America claims the isolated BRCA1 gene as a chemical molecule [31] and certain mutations, the corresponding proteins [32] and antibodies [33] and the conceivable therapeutic applications such as gene therapy, drug screening [34] or a transgenic animal [35], as well as diagnostic kits [36].

3.2.4. EP 0 785 216 – "Chromosome 13-Linked Breast Cancer Susceptibility Gene BRCA2"

The fourth and last patent was granted in January 2003 to Myriad Genetics, Endo Recherche, Inc., Sainte-Foy, Canada, The Trustees of the University of Pennsylvania, Philadelphia, USA and HSC Research and Development Limited Partnership, Toronto, Canada. It covers the sequence of the BRCA2 gene [37], mutations of that gene [38] and method for comparison of the original and the mutated sequence [39].

All in all, the Myriad Genetics patent portfolio covers the gene BRCA1 and BRCA2, specific mutations of these genes, methods for the identification of these mutations for the sake of predictive breast and ovarian cancer diagnostics and kits for the implementation of these methods. The particular importance of these patents is due to the fact that Myriad gained rights on the BRCA genes themselves [40]. This brings about the fact that each technique using the sequence can be prosecuted for patent infringement and every third party using the genes in a following patent application can only achieve a dependent patent.

4. THE CONCERNS

Potential impacts of patenting disease-associated genes on public health care and biomedical research take centre stage in the discussion. As already shown, the case of the BRCA1 genes is not unique because of the growing number of patents on genes, gene sequences and proteins. The case of breast cancer is a significant example for the current major problem of the patenting of disease-associated genes and developed genetic tests [41].

Currently, Myriad owns 260 patents worldwide on BRCA1 and BRCA2, the genes themselves or diagnostic methods based on them [42]. In this way, it is expected that testing will either become impossible in the laboratories in the public sector, or will become very expensive. There is also a certain fear regarding the loss of quality in medical practice and genetic counselling and analysis. Constraints for research and development in the field of molecular genetic predictive breast cancer diagnostics are likely, as Myriad genetics would be able to build up a unique genetic databank of mutations inaccessible to the public. This could finally lead to a slower development or hindrance of new therapies associated with BRCA-related or other tumours.

4.1. Costs of tests

The patents held by Myriad Genetics cover all diagnostic technologies and products that use the sequence of one of the BRCA genes. All laboratories using this sequence could be prosecuted for patent infringement, regardless of whether the technology is used for testing. In this way, Myriad Genetics may enforce that all probes be sent to the United States. The costs for testing would multiply considerably. It is estimated that testing is three and a half times more expensive at Myriad than in France [43], for example. This would, based on current epidemiological data, lead to an increase of the costs by approximately €110 million over 20 years, which corresponds to €5.5 million annually [44]. This simulation could also be extended to other European countries. From these numbers it becomes obvious that this builds a barrier to the access to diagnostic testing and may eventually cause problems for reimbursement systems and negatively influence health care all over Europe. This prediction is supported by the US Study on the effects of patents and licenses on the offering of clinical genetic testing "Effects of Patents and Licenses on the Provision of Clinical Genetic Testing Services" [45]. The participants coincidently agreed that the costs would rise for both the laboratories and the patients [46]. Further evidence exists for hospitals and clinics that patents increase the costs for genetic testing and make the access for patients difficult or impossible [47].

4.2. Quality loss of testing

The monopoly of one firm performing the genetic testing negates independent assessment of quality assurance. In this scenario, there is no possibility for the patient to acquire a second opinion, and a lack of quality control for the performed tests.

The quality of the tests is particularly dangerous as the Myriad tests fail in 10–20% of the cases. Especially, large-size mutations are not detected by industrial methods focusing on the detection of point or small-sized abnormalities. It was ascertained by a group at the Institut Curie [48] that in particular, the direct sequencing method used by Myriad Genetics is unsatisfactory.

4.3. Quality loss of medical practice

A loss of quality in medical practice and genetic counselling and analysis must be expected. Comprehensive and multidisciplinary medical attendance from initial counselling to care would be impossible. Taking into account not only the medical but also the psychological aspects of diagnosis as well as the clinical history of high-risk patients and their families, there would be little or no room for holistic treatment. This would mean that real follow-up of high-risk patients and their families would not be achievable.

Concerns have also been expressed in particular about the provision of medical genetic testing directly to the public, rather than through medical practitioners [49]. This would probably generate genetic testing without an adequate pre- and post-test genetic counselling of the patient and, where needed, also of their family.

4.4. Commercialisation of medical genetic testing

A further concern is that the commercialisation of genetic testing through patents on disease-associated genes, which may encourage the inappropriate marketing and supply of genetic testing services and products [50]. Physicians or scientists may preferably use the test they are financially or scientifically interested in. This might lead to a use of non-reliable or not reliably interpretable genetic test. This means that commercial considerations could also result in genetic tests being offered commercially prematurely, before the results of testing can be properly interpreted, evaluated and used [51].

Attempts to "create markets" for genetic tests, prevention or treatment of doubtful clinical utility might lead to broadening of the definition of disease by simultaneously narrowing the definition of "normal" in the society [52].

In the case of Myriad Genetics, the applied rules for recommendation of susceptibility genetic testing are broader than the ones used by national cancer institutes [53]. Myriad only accepts samples handed in by doctors and not by patients directly and to which an "informed consent" must be included. At the same time, on Myriad's homepage the rules of recommendation are presented in the form of a "quiz" [54] and Myriad recommends that the results of this quiz should be brought to a consultation with a doctor. Actually the pressure that patients can exercise on the doctor should not be underestimated [55]. Additionally, Myriad itself proposes a list of doctors for the genetic counselling [56]. From these recommendations and the extraordinarily aggressive advertising strategy, it is reasonable to assume that this will lead to a more frequent application and belittlement of the genetic testing procedure for breast cancer.

4.5. Further research growth and development of diagnostic methods and new therapies

The separation of biological research, clinical investigation and patient care might result in constraints in research and development in the field of molecular genetic predictive breast cancer diagnostics, as the fruitful communication and cooperation would die away i.e. clinicians often provide relevant patient history and results from earlier investigations to the testing laboratory and in many cases directly to the scientists performing the testing. Moreover, there would be an immense loss of expertise in the field of genetic testing which might probably lead to a loss in funding. The worst issue is one of patient data privacy. By the compulsory sending of samples obtained from high-risk individuals to Myriad Genetics [57], they would be able to build the only genetic databank of mutations that will be inaccessible to the public. This in turn will grant an unchallenged control over the main research materials concerning genes coding for breast and ovarian cancer predisposition, thereby allowing to make further discoveries and ultimately filing further patent applications as a result of such discoveries [58] by other groups highly improbable.

Most probably, this will become a hindrance to new therapies associated with BRCA-related and other tumours. Any gene therapy treatment will require the use of not only a gene carrier or "vector" but also the basic genetic sequence. To potentially cure breast cancer with gene therapy, the first step must be to identify the genetic mutation causing the disease, which might be possible not only for BRCA1 and BRCA2 but also for other hereditary breast cancers. If this gene is patented, however, any gene therapy treatment will depend, at least in part, on the availability of a license from the patent holder [59].

The reach of this licensing problem becomes obvious when one takes into account the fact that all samples obtained from high-risk individuals would be sent to Myriad Genetics. As already explained, this leads to unchallenged control by Myriad Genetics over the main research materials concerning genes coding for breast cancer predisposition. That will subsequently allow Myriad to be the first and maybe the only one to detect and patent new gene mutations. Therefore, gene therapies based on other genes than BRCA1 and BRCA2 would also probably depend on licensing from Myriad Genetics.

Taken into consideration that Myriad might test more and more people independently from their familiar predisposition the problem of the genetic databank of mutations at the exclusive disposal of Myriad Genetics gains even more weight. If more and more people without a familiar predisposition are tested, Myriad might also get uniquely broad information about the reasons for spontaneous cancer and the underlying genetic basis. This could hinder the development of new therapies not only for hereditary cancer but also for spontaneous cancer.

The patent holder could decide whether or not to develop new treatments or prevention strategies, or to develop them more slowly, or to develop them only for some of the potential applications, thereby determining the direction and pace of developments [60].

Finally, many patents which assert rights over human DNA sequences include claims to the use of the sequence for gene therapy, even though such applications have almost never been demonstrated, because patent applicants have been allowed to assert rights over uses which are judged theoretically credible, without having evidence from research to show that they have made experimental progress towards realising this theoretically obvious possibility [61]. In this case, a possibly new genetic therapy would not only depend on the licensing of the patent holder, but could be further prosecuted for patent infringement if it is using the patented sequence.

4.6. Effects of licensing

Regarding this point, empirical data exist in a study [62] carried out by large pharmaceutical companies, biotech start-ups, universities and other publicly funded research institutes and clinical institutions involved in genetic testing. Although there was no general negative effect of patents on genes shown, an impact on the choice of research topics could be demonstrated. Researchers in most cases refrain from research in further uses or functions of a gene once they have realised that it had already been patented by a third party. The perspective of being dependent on a third party's dominant patent, once a commercially usable product has been developed, is clearly disliked [63].

Hence, the patenting of the gene sequence may in fact have a negative influence on the research and development of new therapies.

4.7. Summary

Although the increase of costs is a direct consequence of patenting, patenting creates an important possibility for the amortisation of costs, and therefore an important stimulus for capital investments in research. Myriad Genetics, for example, invested up to $10 million simply for identification and sequencing of BRCA1.

The other concerns presented above however have clearly exposed that there is a threat to access as well as to the development of diagnostic methods and new therapies, even though more quantitative detailed work is needed in this area [64].

In addition, the concerns have shown that the predominant problem of patenting is due to the policy of the patent holder – in the case of BRCA gene, Myriad Genetics – including exclusive licensing and extraordinarily high royalties. In the following paragraphs, the possible ways to tackle the existing problems in Patent Law will be examined, keeping in mind the importance of a patent protection for biomedical research and development.

5. PATENTING CONDITIONS AND OPPOSITION PROCEDURES IN THE CASE OF THE BRCA GENES

All of the four European patents of Myriad Genetics on the BRCA1 and BRCA2 were or are in an opposition procedure at the EPO.

Oppositions to the patent EP 699 754 were filed in October 2001 by a number of parties including the Institut Curie, other French research institutes and various national centres for human genetics [65]. Oppositions to EP 705 902 have been filed in February 2002 by nine parties including again the Institut Curie [66]. Oppositions to EP 705 903 have been filed in August 2002 by six parties including yet again the Institut Curie [67]. Oppositions to EP 0 785 216 have been filed by several parties and once more one of them was the Institut Curie [68].

The oppositions draw on the non-fulfilment of the patenting conditions under the provisions of European Patent Law.

5.1. Patent conditions

As already explained the patent constitutes an exclusive right granted for an invention. An "invention" may be defined as a proposal for the practical implementation of an idea for solving a technical problem. An invention must fulfil the following conditions to be protected by a patent under the EPC: novelty [69], inventive step [70] and an industrial application [71]. It must show an element of novelty, that is, prior to the date of filing, it was not already known to the public in any form and it is not known in the body of existing knowledge in its technical field. This body of existing knowledge is called "prior art". The invention must show an inventive step which could not be deduced by a person with average knowledge of the technical field. It must have industrial applicability, namely it can be made or used in any kind of "industry".

5.2. Opposition procedure

The oppositions draw particularly on the non-fulfilment of the patenting conditions novelty, inventive step and the insufficient description.

5.2.1. Lack of priority and absence of novelty

The first objection is based on an argument, which shows the lack of priority and hence an absence of novelty.

Any gene and the protein coded by this gene are defined by their sequences. The sequences filed by Myriad in 1994 in their first patent applications in the United States were erroneous. Since, as argued by the opponents, the sequences do not correspond to the gene described in the third patent, they are therefore invalid for that patent. Therefore, the reference sequences are no longer those of September 1994, but rather they are those of March 1995. That date then is the real priority date that can be claimed by Myriad Genetics. At that time, however, the BRCA1 gene had been isolated and its correct full sequence was available in the scientific databases. There is therefore a total absence of novelty [72].

Furthermore before Myriad Genetics' patent filing, a number of predisposition tests based on indirect methods were already available [73].

5.2.2. Lack of inventive step

There is also a lack of inventive step. In particular the excessively broad scope is seen as not corresponding to the significance of Myriad's contribution to the public domain, at the date the patent was filed.

At the time the patent was granted, the BRCA1 gene was already located and considerable information regarding the gene and its relevance for the detection of breast and ovarian cancer was already provided. What remained to be done was the final sequencing, a routine operation the results of which should at the most be protected by limited monopoly rights [74].

Myriad Genetics may have won the race to breast and ovarian cancer predisposition genes in the very last stretch, but it did benefit from the knowledge existing in the scientific community [75].

5.2.3. Insufficient description

To this end, the opponents object with the opposition that the invention was insufficiently described in the patent specification. The protein sequence used for Myriad Genetics' first patent filing on diagnostic methods is, *per se*, insufficient for a person with average knowledge of the technical field for producing a predisposition test, as the given gene sequence is erroneous [76].

5.3. Decisions of the opposition division

Public hearings on some of these objections have already taken place and decisions were reached on the basis of these hearings by EPO's opposition division. For one of these decisions, the written statement of grounds for the opposition division's decision is already available. For the others they are not. To date, only one of the decisions has been appealed.

This means the decision process has yet to come to an end. However, it is highly interesting to investigate the decisions ruled so far to see in which direction the future trend is going.

5.3.1. EP 0 699 754

Following a public hearing on 18 May 2004, the opposition division of the EPO decided that the grounds for opposition prejudiced the maintenance of the patent [77]. According to the decision revoking the patent [78] the opponents could show a large number of inadequacies, notably in the gene sequence [79]. In view of the excessive monopoly sought, the Opposition Division rejected the patent and its various additional claims because of lack of inventiveness [80]. In absence of any valid requests by the patent holder it was revoked.

The patent holder decided to appeal against the decision to revoke the patent, but there is no date yet for the next proceeding.

5.3.2. EP 0 705 903

After two days of public hearings on 24 and 25 January 2005, the opposition division has concluded that the patent can be maintained in an amended form [81].

The Opposition Division for specific claims followed the argumentation of the opponents regarding the priority date [82] and novelty [83]. Initially, this patent related to 34 mutations in the BRCA1 gene each of which could be used in testing for breast and ovarian cancer predisposition. The patent was amended at the opposition proceedings and limited the patent to a single mutation [84]. In this way, the patent scope was considerably restricted and now relates only to a gene probe of a defined composition for the detection of a specific mutation; it no longer includes claims for diagnostic methods [85]. The patent therefore will not limit screening for this mutation since identification is possible with other probes differing in length and position from the above [86].

The written statement of grounds for the opposition division's decision is not yet available.

5.3.3. EP 0 705 902

Following public hearing on 19 and 21 January 2005, the opposition division has decided that this patent is also to be maintained in an amended form [87].

The patent claimed the isolated BRCA1 gene, as a chemical molecule and certain mutations, the corresponding proteins and antibodies and therapeutic applications like gene therapy, drug screening or a transgenic animal as well as diagnostic kits [88]. The principal claim concerning the gene itself was rejected as well as the essential points of the other claims, on the grounds of failure to comply with Art. 83 EPC, as it lacked a clear and complete description [89]. Secondary claims concerning a nucleic acid probe and vectors comprising certain gene sequences have been granted [90].

Accordingly, the patent now relates to a gene probe of a defined composition and no longer includes claims for therapeutic and diagnostic methods [91]. Now that the scope of the patent is curtailed in this way, the granted sequences have no impact on diagnostic work and essentially the patent no longer constitutes a hindrance to the implementation of diagnostic tests in research and health care institutions [92].

The written statement of grounds for the opposition division's decision is also in this case not yet available.

5.3.4. EP 0 785 216

Public hearing for the patent EP 0 785 216 at the EPO's Opposition Division is scheduled for 29 June 2005, no decision has been taken so far.

6. CONCLUSION AND OUTLOOK

From the decisions presented above, the trend appears to suggest that some of the objections were well founded. The excessive monopoly of Myriad Genetics has been cut down to a reasonable size, adequate to the contributions of Myriad. Thus, for some of the concerns patent law has already received a convincing answer or, in the words of the Royal Society:

> the best way forward is to tackle the abuse rather than to change the patent law [93]

There is, however, one fundamental point of critique with regards to the EPO's code of practice when dealing with the patenting of disease-associated genes and any gene sequence that should urgently be taken into consideration. Bearing in mind the importance of patent protection for biomedical research and development, the danger of excessive monopoly rights and in succession the danger of a reward for the inventor that is not corresponding to the significance of the contribution to the public domain he has made is especially high in these cases.

Inventions based on genetic sequences have so far been treated like other chemical substances by the EPO. The Patent is not limited to the specified functions of the gene, but is extended to all further possible uses, independently of whether they were known at the date of the patent application or not [94].

Genes, however, differ in various ways from other chemical substances. The most fundamental point is the fact that besides being a chemical substance gene sequences are always a carrier of information. This information, multifunctional in itself, codes for proteins, a few hundred thousands in case of a human body, all of them coded in only 35,000–40,000 genes, but also that some 40% of them are alternatively spliced, i.e. encoding for more than one protein, depending on the combination of exons read in an open reading frame, or even depending on the direction in which exons are read. Beyond this, they can also hybridise with other gene sequences [95]. To cover these multiple possibilities with patent protection by simply identifying the sequence seems insufficient, as the diversity of subsequent implications does not necessarily correspond to the significance of the contribution the inventor has made.

This becomes more obvious if one takes into account the development of sequencing techniques in the past years. It took the Human Genome Project four years to sequence the first billion base pairs, four months to sequence the second billion and subsequent to this 10 (of the three billion base pairs of the entire human genome were sequenced each month [96]. We therefore see that the contribution made through sequencing gets smaller and smaller.

Finally, the EPO's code of practice in gene sequences may hinder the development of new therapies more powerfully than it does for other chemical substances [97]. The possibility of an avoidance of the patent results from the singularity of the gene nearly non-existent.

I would like to specify clearly that I am not against the patenting of genetic information, even human genetic information. There can be no doubt that patents are needed in order to encourage the costly development of useful products from genetic discoveries and to get the information from the secrecy of laboratories or from arcane scientific publications out into the public. But it must be ensured that certain conditions, partly already available in Patent Law, are fulfilled and others are created in other areas of the law, i.e. a Law on genetic testing [98] or in Licensing Law the situation that anyone interested can obtain a licence for further fruitful development.

Given all the needed credit to the stimulating effect of patent law one may not be mistaken when saying that too much legal protection can weaken competition and innovation:

> We often talk about how important patents are to promote innovation, because without patents, people don't appropriate the returns to their innovation activity, and I certainly very strongly subscribe to that. On the other hand, some people jump from that to the conclusion that the broader the patent rights are, the better it is for innovation, and that isn't always correct, because we have an innovation system in which one innovation builds on another. If you get monopoly rights down at the bottom, you may stifle competition that uses those patents later on, and so the breadth and utilization of patent rights can be used not only to stifle competition, but also [can] have adverse effects in the long run on innovation. We have to strike a balance [99].

This balance has to be found in the case of patenting disease-associated genes. In this case, this involves the balance between providing temporary monopoly rights and determining appropriate returns. These returns should correspond to the contribution of the inventors and their innovative creativity and the temporary monopoly rights allotted should allow for free space for the creativity of subsequent inventors to research and develop new drugs and therapies.

REFERENCES AND NOTES

[1] See report "U.S., Britain urge free access to human genome data", available at http://edition.cnn.com/2000/HEALTH/03/14/human.genome03/index.html visited October 17, 2005.
[2] For more details and further reading see Straus J., Patentrechtliche Probleme der Gentherapie, *GRUR*, 1996, **45**, 10 et seqq.
[3] Bisbee, G.E., Will the broad ex vivo gene therapy patent be challenged: An appraisal, *GEN*, 1995, **15**(10), 22 et seqq.
[4] US Patent Nr. 5,399,346 of the inventors W. French Andersin, M. Blaese and S. A. Rosenberg, held by the United States of America (NIH).
[5] Compare Pfeiffer, Sandoz Ltd. Signs Definitive Agreement to Acquire Genetic Therapy for $ 295 M., *GEN*, 1995, **15**(August 14), 1.
[6] Neue Züricher Zeitung 23/24 August 2003, p. 27.
[7] Straus, J., Herrlinger, K.A., Zur Patentierbarkeit von Verfahren zur Herstellung individuum-spezifischer Arzneimittel, *GRURInt*, 2005, **54**, 869, 870.
[8] Evans, J.P., Syrzynia, C., Burke,W., The complexities of predictive genetic testing, *BMJ*, 2001, **322**, 1052, 1054.
[9] US 6,0205,130 "Heriditary Hemochromatosis Gene".
[10] US 5, 712,098 "Heriditary Hemochromatosis Diagnostic Markers and Diagnostic Methods"; US 5,753,438 "Method to Diagnose Heriditary Hemochromatosis" and US 5,705,343 "Method to Diagnose Heriditary Hemochromatosis".
[11] EP 0 954 602 "Heriditary Hemochromatosis Gene".
[12] EP 0 827 "Method to Diagnose Heriditary Hemochromatosis".
[13] US 5,679,635 "Aspartoacylase Gene, Protein, and Methods of Screening for Mutations associated with Canvan Disease".
[14] In the US more than 180 000 cases of breast cancer diagnosed each year; see Kodish E., Wiesner, G.L., Mehlman, M., Murray, T., Genetic Testing for Cancer Risk: How to Reconcile the Conflicts, *JAMA*., 1998, **279**, 179.
[15] McPherson, K., Steel, C.M., Dixon, J.M., ABC of breast diseases. Breast cancer-epidemology, risk factors and genetics, *BMJ*, 1994, **309**, 1003; Broeders, M.J., Verbeek, A.L., Breast cancer epidemiology and risk factors, *Q J Nucl Med*., 1997, **41**, 179.
[16] Miller, A.B., Bulbrook, R.D., UICC multidisciplinary project on breast cancer: The Epidemiology, Aotiology and Prevention of Breast Cancer, *Int. J. Cancer*, 1986, **37**, 173.
[17] Agnarsson, E., Pennisi, E., Roberts, L., In the Crossfire: Collins on Genomes, Patents, and Rivalry, *Science*, 2000, **287**, 2369.
[18] Given that they reach the age 85.
[19] Hall, M., Lee, M.K., Newman, B., Morrow, J.E., Anderson, L.A., Huey, B., King, M.C., Linkage of early-onset familial breast cancer to chromosome 17q21, *Science*, 1990, **250**, 1684 et seq.
[20] Miki, Y., Swensen, J., Shattuk-Eidens, D., Futreal, P.A., Harshman, K., Tavtigian, S., Liu, Q., Cochran, C., Benett, L.M., Ding, W., A strong candidate for the breast and ovarian cancer susceptibility gene BRCA, *Science*, 1994, **266**, 66; Futreal, P.A., Liu, Q., Shattuck-Eidens, D., Cochran, C., Harshman, K., Tavtigian, S., Bennett, L.M., Haugen-Strano, A., Swensen, J., Miki, Y., Eddington, K., McClure, M., Frye, C., Weaverifeldhaus, J., Ding, W., Gholami, Z., Soderkvisi, P., Terry, L., Jhanwar, S., Berchuck, A., Iglehart, J.D., Marks, J., Ballinger, D.G., Barrett, J.C., Skolnick, M.H., Kamb, A., Wiseman, R., BRCA1 Mutations in Primary Breast and Ovarian Carcinomas, *Science*, 1994, **266**, 120.
[21] Marx, R., A second breast cancer susceptibility gene is found, *Science*, 1996, **271**, 30; Gitter, D.M., International Conflicts over patenting human DNA sequences in the United States and the European Union: An argument for compulsory licensing and a fair use exeption, *N.Y.U. Rev*., 2001, **76**, 1623, 1652.

[22] Wooster, R., Bignell, G., Lancaster, J., Swift, S., Seal, S., Mangion, J., Collins, N., Gregory, S., Gumbs, C., Micklem, G., Barfoot, R., Hamoudi, R., Patel, S., Rices, C., Biggs, P., Hashim, Y., Smith, A., Connor, F., Arason, A., Gudmundsson, J., Ficenec, D., Kelsell, D., Ford, D., Tonin, P., Bishop, D.T., Spurr, N.K., Ponder, B.A.J., Eeles, R., Peto, J., Devilee, P., Cornelisse, C., Lynch, H., Narod, S., Lenoir, G., Egilsson, V., Barkadottir, R.B., Easton, D.F., Bentley, D.R., Futreal, P.A., Ashworth, A., Stratton, M.R., Localisation of a Breast Cancer Susceptibility Gene, BRCA2, to Chromosome 13q12–13, *Science*, 1994, **264**, 2088 et seqq, see Miki, Y., Swensen, J., Shattuk-Eidens, D., Futreal, P.A., Harshman, K., Tavtigian, S., Liu, Q., Cochran, C., Benett, L.M., Ding, W., A strong candidate for the breast and ovarian cancer susceptibility gene BRCA, *Science*, 1994, **266**, 66: "A second locus, BRCA2, recently mapped to chromosome arm 13q (6), appears to account for a proportion of early-onset breast cancer roughly equal to that resulting from BRCA1".
[23] Marx, R., *Science*, 1996, **271**, 30.
[24] Published in Wooster, R., Bignell, G., Lancaster, J., Swift, S., Seal, S., Mangion, J., Collins, N., Gregory, S., Gumbs, C., Micklem, G., Barfoot, R., Hamoudi, R., Patel, S., Rices, C., Biggs, P., Hashim, Y., Smith, A., Connor, F., Arason, A., Gudmundsson, J., Ficenec, D., Kelsell, D., Ford, D., Tonin, P., Bishop, D.T., Spurr, N.K., Ponder, B.A.J., Eeles, R., Peto, J., Devilee, P., Cornelisse, C., Lynch, H., Narod, S., Lenoir, G., Egilsson, V., Barkadottir, R. B., Easton, D. F., Bentley, D. R., Futreal, P. A., Ashworth, A., Stratton, M. R., Identification of the breast cancer susceptibility gene BRCA, *Nature*, 1995, **378**, 789 et seqq.
[25] US 5,709,999, S. 1.
[26] Claims 1–3.
[27] Claims 3–6.
[28] Claims 7 and 8.
[29] Claim 14.
[30] Claim 16.
[31] Claims 1–4.
[32] Claims 15–18.
[33] Claim 19.
[34] Claims 27–32
[35] Claim 34.
[36] Claims 25 and 26.
[37] Claim 1.
[38] Claims 3–6.
[39] Claim 15.
[40] Herrlinger, K.A., Die Patentierung von Krankheitsgenen, 2005, p. 72.
[41] *Ibid.*, p. 7.
[42] *Ibid.*
[43] "Initial family mutation searches performed by Myriad are billed 2,400! dollars (18,000 francs – 2,744 euros), as against an estimated cost of 5,000 francs (762! euros) for testing in French laboratories, which makes Myriad three and a half times more expensive."; see *http://www.curie.fr/upload/presse/myriadopposition6sept01_gb.pdf* visited October 17, 2005.
[44] see *http://www.curie.fr/upload/presse/myriadopposition6sept01_gb.pdf* visited October 17, 2005.
[45] For detailed description of the study see Cho, M.K., Illangasekare, S., Weaver, M.A., Leonard, D.G.B., Merz J.F., Effects of Patents and Licenses on the Provision of Clinical Genetic Testing Services, *J. Mol. Diagn.*, 2003, **5**, 3 et seqq.; for discussion of the critics on the study see Herrlinger, p. 165 et seqq.
[46] Cho, M.K., Illangasekare, S., Weaver, M.A., Leonard, D.G.B., Merz J.F., Effects of Patents and Licenses on the Provision of Clinical Genetic Testing Services *J. Mol. Diagn.*, 2003, **5**, 3, 6.

[47] Chahine K., Industry opposes genomic legislation, *Nat. Biotechnol.*, 2002, **20**, 419.
[48] Gad, S., Scheuner, M.T., Pages-Berhouet, S., Caux-Moncoutier, V., Bensimon, A., Aurias, A., Pinto, M., Stoppa-Lyonnet, D., Identification of a large rearrangement of the BRCA1 gene using colour bar code on combed DNA in an American breast/ovarian cancer family previously studied by direct sequencing, *J. Med. Genet.*, 2001, **38**(6), 388–391.
[49] See Australian Law Reform Commission and Australian Health Ethics Committee, Essentially Yours: The Protection of Human Genetic Information in Australia, ALRC 96 (2003), ALRC, Sydney, see www.alrc.gov.au visited October 17, 2005, Chapter 11; Human Genetics Society of Australasia, HGSA Position Paper on the Patenting of Genes (2001), HGSA, see www.hgsa.com.au/policy/patgen.html visited October 17, 2005.
[50] http://www.austlii.edu.au/au/other/alrc/publications/issues/27/12._Gene_Patents_and_Healthcare_Provision. doc.html visited October 17, 2005.
[51] Ontario Ministry of Health and Long-Term Care, Genetics, Testing & Gene Patenting: Charting New Territory in Healthcare — Report to the Provinces and Territories (2002), Ontario Government, available at www.gov.on.ca visited October 17, 2005, 44.
[52] Human Genetics Society of Australasia, HGSA Position Paper on the Patenting of Genes (2001), HGSA, see www.hgsa.com.au/policy/patgen.html visited October 17, 2005.
[53] Browe V., Testing, testing… testing? *Nat. Med.*, 1997, **3**, 131; Smith, O., Dissension greets breast cancer panel consensus, *Nat. Med.*, 1997, **3**, 709; Caulfield, T.A., Gold, E.R., Genetic testing ethical concerns, and the role of patent law, *Clin. Genet.*, 2000, **57**, 370–371.
[54] See "the hereditary cancer quiz" on http://www.myriadtests.com/quiz.htm?s=View visited October 17, 2005.
[55] There exists empirical evidence from a study that 95% of all women would undergo a genetic test in opposition to their doctors' advice, see Caulfield T.A., Gene testing in the biotech century: Are physicians ready? *JAMC*, 1999, **161**, 1222.
[56] See *http://www.myriadtests.com/finddoc.htm* visited October 17, 2005.
[57] See *http://www.curie.fr/upload/presse/190504_fr.pdf* visited October 17, 2005.
[58] See *http://www.curie.fr/upload/presse/myriadopposition_feb02_gb.pdf* visited October 17, 2005.
[59] See *http://www.austlii.edu.au/au/other/alrc/publications/issues/27/12. _Gene_Patents_and_Healthcare_Provision.doc.html* visited October 17, 2005.
[60] See *http://www.hgsa.com.au/policy/patgen.html* visited October 17, 2005, 2.2–2.4.
[61] Nuffield Council on Bioethics, The Ethics of Patenting DNA (2002), Nuffield Council on Bioethics, London, available at www.nuffieldbioethics.org visited October 17, 2005, 61. See the discussion of 'usefulness' in Chapter 9.
[62] Straus, J., Holzapfel, H., Lindenmeir, M., Empirical Study on "Genetic Inventions and Patent Law", Munich, 2002.
[63] Straus, J., Holzapfel, H., Lindenmeir, M., Empirical Study on "Genetic Inventions and Patent Law", Munich, 2002, No 6c.
[64] Herrlinger, K.A., Die Patentierung von Krankheitsgenen, 2005, p. 188.
[65] Other opponents include: Assistance publique – Hôpitaux de Paris, the Institut Gustave Roussy, the Belgian Society for Human Genetics and the Associazione Angela Serra per la Ricerca sul Cancro.
[66] Opponents besides of the Institut Curie: The Social Democratic Party of Switzerland, Berne; Greenpeace Germany, Hamburg; Assistance publique – Hôpitaux de Paris, Paris; the Institut Gustave Roussy, Villejuif (F); the Belgian Society of Human Genetics *et al.*, Brussels; Dr Wilhelms, Göhrde (D); the Netherlands, represented by the Ministry of Health, The Hague; and the Austrian Federal Ministry of Social Security, Vienna

[67] Opponents besides of the Institut Curie: Assistance publique – Hôpitaux de Paris; the Institut Gustave Roussy; the Vereniging van Stichtingen Klinische Genetica, Leiden (NL); the Netherlands, represented by the Ministry of Health; and Greenpeace Germany.
[68] Other opponents include: The Belgian Society of Human Genetics *et al.*, Brussels; Assistance publique - Hôpitaux de Paris, Paris:
[69] Art. 54(1) EPC.
[70] Art. 56 EPC.
[71] Art. 57 EPC.
[72] *http://www.gene.ch/genet/2002/Oct/msg00007.html* visited October 17, 2005.
[73] *http://www.curie.fr/upload/presse/myriadopposition6sept01_gb.pdf* visited October 17, 2005.
[74] *http://www.gene.ch/genet/2002/Oct/msg00007.html* visited October 17, 2005.
[75] *http://www.curie.fr/upload/presse/myriadopposition6sept01_gb.pdf* visited October 17, 2005.
[76] *http://www.curie.fr/upload/presse/myriadopposition6sept01_gb.pdf* visited October 17, 2005.
[77] See press release of the EPO *http://www.european-patent-office.org/news/pressrel/2004_05_18_e.htm* visited October 17, 2005.
[78] All documents available at http://my.epoline.org/portal/public visited October 17, 2005.
[79] See EPO, on patent EP 0 699 754, Decision revoking the European Patent (Article 102(1) EPC), document from 3.11.2004, 5 et seqq.
[80] EP0, on patent EP 0 699 754, Decision revoking the European Patent (Article 102(1) EPC), document from 3.11.2004, 22; or *http://www.curie.fr/upload/presse/myriadpatents310105.pdf* visited October 17, 2005.
[81] See press release of the EPO *http://www.european-patent-office.org/news/pressrel/2005_01_21_e.htm* visited October 17, 2005.
[82] See EPO on patent EP 0705903, Minutes of the oral proceedings (Opposition division) - introduction of the parties, document from 25.1.2005, 3.
[83] See EPO on patent EP 0705903, Minutes of the oral proceedings (Opposition division) - introduction of the parties, document from 25.1.2005, 4.
[84] *http://www.curie.fr/upload/presse/myriadpatents310105.pdf* visited October 17, 2005.
[85] See press release of the EPO *http://www.european-patent-office.org/news/pressrel/2005_01_21_e.htm* visited October 17, 2005.
[86] *http://www.curie.fr/upload/presse/myriadpatents310105.pdf* visited October 17, 2005.
[87] See press release of the EPO *http://www.european-patent-office.org/news/pressrel/2005_01_25_e.htm* visited October 17, 2005.
[88] See above 7 et seq. for more detailed information.
[89] See EPO on patent EP 0705902, Minutes of the oral proceedings (Opposition division) - introduction of the parties, document from 19.1.2005. 4.
[90] *http://www.curie.fr/upload/presse/myriadpatents310105.pdf* visited October 17, 2005.
[91] See press release of the EPO http://www.european-patent-office.org/news/pressrel/2005_01_25_e.htm visited October 17, 2005.
[92] http://www.curie.fr/upload/presse/myriadpatents310105.pdf visited October 17, 2005.
[93] The Royal Society, Keeping science open: the effects of intellectual property policy on the conduct of science, 10, available at *http://www.royal-soc.ac.uk/document.asp?tip=0&id=1374* visited October 17, 2005.
[94] See Herrlinger, K.A., Die Patentierung von Krankheitsgenen, 2005, for a detailed discussion of this problem, p. 133.
[95] Straus, J., special Edition Official Journal of the EPO, 166, available at *http://www.european-patent-office.org/epo/pubs/oj003/07_03/se2_07_03.pdf* visited October 17, 2005; for further detailed information see Venter *et al.*, *Science*, 2001, **291**, 1304, 1346 or Malakoff, D., Will a Smaller Genome Complicate Patent Case? *Science*, 2001, **291**, 1194.

[96] Marshall, E., Pennisi, E., Roberts, L., In the Crossfire: Collins on Genomes, Patents, and Rivalry *Science*, 2000, **287**, 2396.
[97] Markl, H., Who Owns the Human Genome? What Can Ownership Mean with Respect to Genes? IIC, 2002, **33**, 1, 4.
[98] Such a law is discussed right at the moment in Germany the so-called "Gendiagnostik Gesetz".
[99] Stiglitz, G., Address to the U.S. FTC, Hearings on global Competition and Innovation, FTC Report, Chapter 6, 6.

On the Political Side of Gene Therapy, What can be Drawn from the French Situation

Anne-Sophie Paquez

Sociology and Public Policy Program, Institut d'Etudes Politiques (Sciences Po), Paris

Abstract

When the daily *Le Monde* announced on 31 January 1990 that "two patients suffering from cancer were involved in a gene therapy clinical trial" [1], it seemed a new era for medicine was dawning. Applied to cancer, one of the major plagues of our time, gene therapy seemed a miracle. Its process was explained to the public through simple words, on the pattern of a syllogism: given that cancer is caused by a defeated gene and that gene therapy allows one to modify the genome, gene therapy should fight cancer. It sounded logical. Contrary to some other scientific news, more complex to explain to lay people, everybody could see the benefit they could obtain from gene therapy. Consequently, this "manipulation" of the human genome was perceived as positive and received wide support from the French public. According to the politicians, gene therapy was "without any contest, a fantastic future path" [2] to enhance public health. It would also affirm French "grandeur" to the world, placing France among the international leaders in the competitive field of biomedical research and industry.

Beside these radiant hopes, the tragic hazards intrinsically linked with gene therapy were under public discussion. Modifying the human genome, this most private and basic component of human beings, via gene therapy was not a common medical practice. In a context of fast-growing application of biotechnology to human beings [3], gene therapy was framed as a political stake. There was no harsh controversy on the opportunity of authorising or banning it. Instead, a consensus was met with the only dilemma being how to set limits to gene therapy, implying to put an end to the principle of the freedom of research. For political leaders, this was a new challenge to tackle, as it broke the tacit rule that the State would not interfere with science regulation. How came, then, the French legislator to regulate gene therapy?

Gene therapy will first be examined in the French societal context, where it is, applied on the *soma*, widely supported by public opinion, thanks to an enthusiastic popularisation by the media and a strong backing from the political authorities (see Section 1). This will lead us to identify the main political dilemmas gene therapy raised and how they were handled. We will stress, here, the impact of gene therapy as a "public problem" on the French traditional policy-style (see Section 2). Broadening our scope, we will, finally, try to sketch the prospective political stakes of gene therapy (see Section 3).

Keywords: lay people, controversy, freedom of research, science regulation, societal context, popularisation, public problem, policy-style, prospective political stakes.

Contents

1. Gene therapy in the French societal context 182
   1.1. Support by the public opinion 182

1.2. Enthusiastic popularisation by the media 183
1.3. A favourable political framework 185
2. The political stakes of gene therapy: Towards a new French policy-style 187
2.1. Coping with sanitary risk 187
2.2. Ensuring morals: Is gene therapy "morally correct"? 189
2.3. Reviewing the French traditional policy-style 191
3. Broadening the scope: gene therapy from a prospective political point of view 192
3.1. Towards a global binding ethical framework? 192
3.2. On the societal side: Towards a social claim for authorising Germ-line gene therapy? 194
3.3. Towards a participatory democracy? 195
4. Conclusion 196
References and Notes 196
Further reading 198
Extracts from Interviews 200

1. GENE THERAPY IN THE FRENCH SOCIETAL CONTEXT

To set the societal scene for gene therapy in France, we will initially focus our discussion on somatic cell gene therapy; the stakes related to germ-line gene therapy will be further investigated (Section 3.2). The situation may be described by three main characteristics: wide support by the public opinion, an enthusiastic popularisation by the media and a favourable political framework.

1.1. Support by the public opinion

Somatic cell gene therapy is chiefly perceived as a "good" biotechnology in France. This is not a specifically French feature, as the 1996 and 2000 Eurobarometer surveys on the social acceptability of biotechnology [4] revealed. In most European countries, the "red" biotechnology is all the more supported as the "green" one is rejected. In 2000, 89% of the French thought it "useful" to "use genetic tests to detect some diseases we might have inherited from our parents"; 79%, to "clone human cells or tissues to replace those who do not work properly"; and 77%, to "introduce human genes in bacteria to produce medicines or vaccines"; whereas only 39%, to "use modern biotechnology in food production"[5]. It is worth pointing out that gene therapy appears to be one of the least controversial "red" biotechnology (when compared to pre-implantation diagnosis or human cloning).

The French are, however, cautious on opening this Pandora's box. In 2001, 75% of them agreed on the fact that science had "developed too fast comparing the morality of human beings" and a much stronger part (83%) thought that "limits should be put to science, since it would go too far on modifying Nature". This attitude seems to have more to do with awareness than an outright rejection: for a prominent majority believes (67%), "science should go on, even if some research projects may disrupt moral principles" [6].

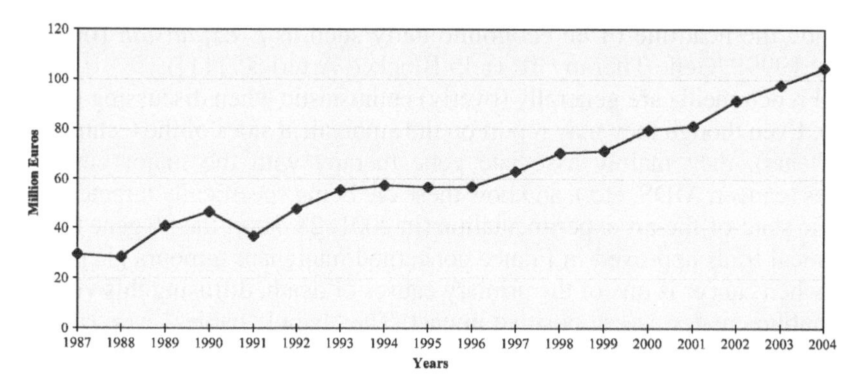

Fig. 1. The "Téléthon" donations (1987–2004).

In a traditional context, as part of general support for medical research in France, gene therapy is considered as progress in scientific knowledge and in medical treatments. In a survey published in January 2001 [7], 88% of the interviewed people wished the discoveries of genetic research would be used to "cure serious diseases". At the same time, in another poll [8], 90% of the interviewees deemed scientific research to be a national priority and 54% that the public expenditure spent on biotechnology research should increase.

The announcement in October 2002 that the gene therapy clinical trial on "bubble babies" had not been as successful as expected did not seem to impinge on this positive attitude. The donations to the "Téléthon" seem to indicate this. From 2002, the donations have continued to increase annually, reaching a sum of 104 million Euros in 2004 (see Fig. 1).

To explain this positive attitude, we may refer to the psychologist analysis of risk [9], which is based upon the idea of risk as a social construct. The social acceptability of risk will then depend on whether people may clearly perceive their own benefits at it. There is a positively correlated relationship between the identification of the goal of a technology and the acceptability of risk. This is the case for gene therapy, since its goal should be to fight diseases, especially cancer, one of the most tragic plagues of our time. This is the predominant message conveyed by the French media.

1.2. Enthusiastic popularisation by the media

The French media displayed a major interest in gene therapy, contributing to set a cognitive framework. Dr. Marina Cavazzana-Calvo, one of the investigators (with Pr. Alain Fischer) of the gene therapy clinical trial on "bubble babies", explained to us [10] that gene therapy was nothing special technically, but was under a high media attention. Gene therapy has been covered by all types of media and, for the press, by all the papers, both elite and popular, generalist and specialised. It was significant that gene therapy had made the headline of an economic daily such as *L'Expansion* (on 21 January 1999: "Gene Therapy To Help Blocked Arteries" [11]).

The French media are generally (overly) enthusiastic when discussing gene therapy. Even though they may report on the ambivalent sides of the technique (hopes/fears), they mainly associate gene therapy with the major current scourges (cancer, AIDS, etc.), and how these are being specifically targeted by scientific state-of-the-art experimentation (in 2001, 28 out of the 36 gene therapy clinical trials approved in France concerned malignant tumours [12]). At a time when cancer is one of the primary causes of death, diffusing this vision to the public creates a very positive impact. The "bubble babies" case raised deep emotions, which the media then exploited. We may observe that the babies were not in the media. Their names were kept secret and there were no photos of them. Only scientific data about the clinical trial were published. Thus, for the French public opinion, gene therapy could not be associated with any pictures, which generally enhance emotions (process of the "mediatisation"). This differs from the United Kingdom, where Rhys Evans, the first "bubble baby" saved by Adrian Thrasher's team at the GOSH [13] in London, is a popular figure (pictures of him in the press and TV broadcasts).

A specific TV broadcast is dedicated to gene therapy: the "Téléthon". This 30-hour programme, which has been broadcast on a public channel every year on the first weekend of December since 1987 (just after the discovery of the gene involved in the Duchenne Myopathy) is the longest live TV show in the world. Adapted from an American idea, introduced by the French Association against Myopathies (AFM), a patients' association aiming at defeating neuromuscular diseases, the "Téléthon" was implemented to raise funds to support the AFM's research programmes on neuromuscular diseases (the AFM is the major actor on genetics research in France) and to inform the public opinion about rare genetic disorders.

The "Téléthon" is much more than an entertainment programme: it is a scientific broadcast on genetics. Patients and scientists (only those whose research projects are funded by the AFM) play a major role in the programme by educating the public. Through the "Téléthon", science enters households. It has become a very popular broadcast and, to some extent, a traditional national event. In 2003, 5 million people participated in the 22,000 local events, which were organised across France. The public audience has reached high rates (see Fig. 2).

The success of the "Téléthon" relies on an understandable discourse, based on the – apparent – simplicity of gene therapy and on the credo of "all genetics". Whether it is scientifically true or not, it is of no doubt that the concept of gene therapy has been brought to the attention of the French society via the "Téléthon" and is positively perceived. D. Kahneman and A. Tversky pointed to the role played by heuristics in social acceptability of risk [14]. According to them, we are used to reducing the complexity of information by resorting to two main "simplificatory" tools: the "heuristics of representation" (a causal process, partly based on stereotypes) and the "heuristics of past" (role played

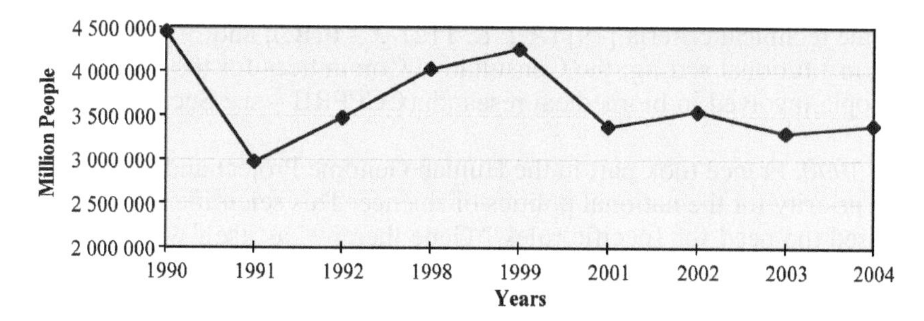

Fig. 2. The "Téléthon" audience (1990–2004).

by our memories). The "Téléthon" may be considered as a tool of the "heuristics of representation".

This overall optimistic coverage of gene therapy by the media has framed the French risk perception. As P. Slovic, B. Fischhoff and S. Lichtenstein wrote [15] that the media play a major role in both risk perception and risk assessment. People tend to think that flying by plane is riskier than smoking because pictures of an airplane crash are much more striking for them than a single person dying from throat cancer. This results from the mediatisation of events. Consequently, people would over-estimate the first risk and under-estimate the second. Besides the media, gene therapy has also been under particular attention from the political leaders.

1.3. A favourable political framework

Since the end of the 1970s, French governments have growingly been supportive of genetic research. By setting incentive legislative rules to favour the development of somatic cell gene therapy and by making genetics a priority for the politics of science, political leaders have conveyed to their citizens the "vision" that somatic gene therapy is safe and morally acceptable.

The French laws on gene therapy consist of a complex, step-by-step built network (intertwined with the European directives). Four main steps may be identified by year of implementation: 1988, 1994, 1996 and 2004.

In 1988, the law on the protection of people involved in biomedical research [16] set the rules for the experiments on human beings. It was reviewed several times, especially in 2001, to transpose the EC directive on clinical trials [17] but its core was not subsequently modified. It currently regulates somatic cell gene therapy trials.

The main points of the 1988 law are:

- the basic conditions to conduct a biomedical research on human beings (Art. L. 1121-2, Public Health Code – PHC);
- a free and express consent (after getting informed [18]) of the person subject of this experiment (Art. L. 1121-2 – PHC);

- some technical criteria [19] (Art. L. 1121-3 – PHC); and
- an institutional setting: the Consultative Committees for the protection of people involved in biomedical research (CCPPRB – see Section 2.1).

In 1990, France took part in the Human Genome Project and made genetics a priority for the national politics of science. This scientific advancement stressed the need for specific rules. "Gene therapy" as itself was put on the political agenda as an issue among those, which had been raised by the late progress of biomedicine and was mentioned in the laws in 1994, under a bioethical frame. In July 1994, the three French laws on bioethics were passed: the first [20] was on the protection of personal data; the second [21] set, for the first time, the basic principles regarding respect of the human body; and the third [22] regulated some taxing issues including Assisted Reproductive Techniques and organ transplants. Gene therapy was mentioned in these latter two laws.

Two main ethical principles amended the Civil Code [23]: "human dignity" and the "integrity of human body". The law stipulated that "attempting at the integrity of the human body is forbidden, except for a special circumstance, the therapeutic necessity for the person" (Art. 16.3). Moreover, "no one shall make an attack at the integrity of human species.

Any eugenic practice leading to the organisation of the selection of people is prohibited. Without any prejudice for research aiming at preventing and treating genetic diseases, no change can be brought to genetic characteristics in a bid to modify the descent of the person" (Art. 16-4). This explicitly forbade germ-line gene therapy and, without expressly saying so, authorised somatic cell gene therapy.

The third law [24] fixed a specific regulatory regime for gene and cell therapy products. Gene therapy products were defined as those "aiming at transferring genetic material [25]", different from cell therapy products [26]. As these regimes have been modified since then, we are not presenting them here (see Section 2.1.).

Meanwhile, the expert group on gene therapy settled by the Minister of Health and supervised by J.-P. Cano and A. Fischer recommended that the gene therapy regulatory regime should be clarified to support its development (when the first national plan on gene therapy was about to be launched). Following this, and as the first French gene therapy clinical trial was about to be launched, the law on "measures concerning the sanitary, social and statutory order", passed on 28 May 1996, [27] gave some details about the status of gene therapy products by inserting a new title, "Gene and Cell Therapy Products", in the Public Health Code. These products were assimilated to the classical medicines' regime, with some exceptions.

This law also settled a "High Council for Gene and Cell Therapies" [28] (Art. 19) (not implemented yet), placed under the authority of the Prime Minister, in charge of presenting him/her advice to develop gene and cell therapies and of co-ordinating the action of public and private bodies involved in these fields.

Gene therapy went out of the political agenda. The 1998 law on "sanitary security" lead to the setting of the French Agency for the Sanitary Safety of Health Products (Afssaps), in charge of assessing and controlling all the medical products applied to human beings. Gene therapy was not at stake when the 1994 laws on bioethics were reviewed to fit with scientific progress [29]. In fact, parliamentary debates focussed on highly controversial issues: human cloning and research on embryos (embryonic stem cells). The new Agency of Biomedicine, in charge of embryology, transplants and genetics, should not play any role in the regulation of gene therapy.

This leads us to identify the main political dilemmas gene therapy raised and how they were handled in France.

2. THE POLITICAL STAKES OF GENE THERAPY: TOWARDS A NEW FRENCH POLICY-STYLE

The passage from a scientific issue to a political matter implies a "publicisation" process: transforming an activity executed by experts in a confined – almost secret – place (the laboratory) into a topic which should be discussed on the public scene (the "agora") by lay people. Gene therapy raised three main political issues: coping with sanitary risk, ensuring morals and, beyond this, reviewing the French traditional policy-style.

2.1. Coping with sanitary risk

Gene therapy, as a new-born technique, is still at the experimental stage and, thus, raises the issue of coping with sanitary risk. It is, for example, unsure whether the targeted tissue would be reached and whether the expected effect would result. The pressure on politicians for sanitary security was particularly strong because of the general context. In the 1990s, many successive scandals (contaminated blood, dioxin, BSE, etc.) prompted distrust by the public opinion towards political leaders and scientific experts. The French legislative framework on gene therapy was highly influenced by this pressure to ensure the most secure conditions for medical research.

In 1996, the legislator wanted to insert gene therapy into the classical regimes regulating medicines or medical trials. However, the regulatory framework finally established in 1999 (in the context of the BSE crisis) created very stringent norms for gene therapy. Adopting a precautionary approach, the legislator implemented a specific regime for gene therapy, based on the idea that the transgene was not a usual medicine because of its genetically modified nature.

We may first notice that the health and research centres involved in gene therapy have to respect very strict rules. Moreover, the legislative framework made a difference between two kinds of activities: preparation, storage, distribution and cession, on the one hand; deduction and administration on the other. These latter two are considered as care activities and subject to specific rules (licensed by the Minister of Health, Art. L. 676-6, Public Health Code).

For the gene therapy products (see Fig. 3), a distinction is drawn according to the quantitatively assessed impact. For medicines or industrial-scale products, the authorisation procedure is simpler: the (dis)approval is granted by the Afssaps. Otherwise, the decision is made by the Minister of Health, in accordance with advice from the Afssaps (Art. L. 676-3 PHC). Here again, a specific process was introduced by the law for the decision to be made by the Afssaps: General-Director (GD) has to consult to a special Commission. As the decrees which should have fixed the membership and the functioning of this Commission have not been published yet, the GD asks two existing commissions: the Genetic Engineering Commission [30] and the Biomolecular Engineering Commission [31].

For the gene therapy clinical trials, a preliminary double-step authorisation, by the CCPPRB then by the Afssaps (see Fig. 3) was made compulsory.

The "investigator" [32] of the research has to submit his/her project to the CCPPRB responsible for in the region where he/she is working (Art. L. 1123-6 PHC). The CCPPRB delivers advice on scientific [33] and ethical grounds within 5 weeks from the reception of the project. Even though the advice of the CCPPRB is non-binding (it is not a decision), this first step is a mandatory stage [34] for the investigator to launch a clinical trial.

Activity	Place of Execution	Authorities in Charge of Control
Deduction of cells	Specifically authorised health or blood transplant centres	Regional Clinical Agency (« Agence régionale d'hospitalisation » - ARH)
Preparation, storage, distribution and cession of gene therapy products	Specifically authorised centres	French Agency for Sanitary Security of Health Products (Agence française de sécurité sanitaire des produits de santé - Afssaps)
Administration of gene therapy products	Specifically authorised health or blood transplant centres	Regional Clinical Agency (ARH)
Clinical trials	Specifically authorised health or blood transplant centres	French Agency for Sanitary Security of Health Products (Afssaps)

+

Authorisation of the product and the clinical trial by the Afssaps

Fig. 3. The French regulatory framework for gene therapy.

In a second step, the "promoter" has to send the Afssaps a cover letter presenting the main points of the research, and the advice from the CCPPRB. The Afssaps should respond within three months.

It is worth noticing here that the Afssaps stands at the centre of the system. This independent agency plays both an upstream role, assessing the products and the protocols, and a downstream role, through its police powers (control). We may also point out that this framework facilitates communication between scientists and the public administration, on the one hand, and between the public administration and the citizens on the other (the Afssaps is also in charge of informing people on risks related to public health).

The second main political dilemma raised by gene therapy was the moral issue.

2.2. Ensuring morals: Is gene therapy "morally correct"?

By giving man the divine and sacred power to modify the basic nature of human beings, of interfering with fate, gene therapy has allowed infinite scope for human action and disrupts the traditional conceptual limits upon which our civilisation was based. Science challenges these limits (their determinism and the way they were set). This leads to new questions being raised, which are intrinsically political (prompting us to think of the social cohesion), with a crucial and original impact.

In trying to answer the question "to what extent could the human genome be modified?", which limits to put to scientific research and medical practice needed to be defined. Gene therapy provided a new context for scientists, who had until then been considered as working for the "good" of mankind and who had never been, in France, at the mercy of such legislation at such an upstream level of their work. The freedom of research was not set in laws but, under a tacit agreement, it was related to the freedom of speech and thought, which were themselves, mentioned in the Universal Declaration of Human Rights, passed under the French Revolution (1789).

But times have changed, and as Hans Jonas wrote: "genetic manipulation: this ambitious dream of the homo faber which is summed up in the motto that man wants to handle his/her own evolution, in a bid not only to preserve the integrity of human species but to improve and to modify it according to his/her own project". For the philosopher, by playing such a "demoniac role", "our power of action brings us beyond the concepts of any traditional ethics" [35]. This ability to make utopia reality, what Lucien Sfez called the "biotechnological dream" [36], has created deep fears in the public. By "fabricating" men, science is no longer fiction. In the 20th century some governments had tried to build a new social order, based on an eugenic vision. Genetics became discredited and this is the main reason why a moral dilemma emerged from gene therapy.

To complete the summary of the cultural and moral background in which the French political leaders were embedded, we may add the issue of the status of the human embryo (since, in germ-line gene therapy, genetic flaws are corrected in the DNA of externally fertilised human embryos), which has been a taboo since the heated debates on the abortion, in the 1970s (1975–1979). The bioethics laws prohibited any experiment, research or manipulation of embryos [37] but did not give a clear definition of the status of the embryo, which left the door open for controversies.

The stake was, for the French legislator, to create a framework under which gene therapy could appear as a "morally correct" technique. Though the political leaders insisted upon the need for preserving morals from science, analysing the legislative decisions leads us to think that they rather tried to defend science against morals.

As we saw before, the French legislator drew a line between germ-line and somatic cell gene therapies. These words were not written by themselves in the legal framework. Furthermore, somatic cell gene therapy was only implicitly authorised, as it was not banned. Article 16-4 of the Civil Code stipulates that "no one shall make an attack at the integrity of human species. Any eugenic practice leading to the organisation of the selection of people is prohibited. Without any prejudice for research aiming at preventing and treating genetic diseases, no change can be brought to genetic characteristics in a bid to modify the descent of the person". We may notice here the reference to eugenics. In the end, however, the (tacit) principle of the freedom of research is saved by an exception for the sole genetic research.

By doing so, the French legislator settled a political boundary based on a scientific basis: the distinction between somatic and germ-line cells (which is, in fact, not so sure since, for some scientists, a somatic cell gene therapy might impact on germ cells). However, the main objective is reached: setting a rational (at least apparently) limit, which could, by the public opinion, legitimate the political choice. In the French context, this also preserves from defining a status for human embryos.

This position followed the guidelines adopted by the international organisations and the French National Consultative Committee on Ethics (CCNE).

In the 1980s, the Parliamentary Assembly of the Council of Europe handed down two recommendations advocating the human "right to a genetic inheritance which has not been artificially interfered with, except in accordance with some human rights respect" [38]. It explicitly forbade "any form of therapy on the human germinal life" [39].

In 1986, the French CCNE suggested to prohibit any "artificial modification, through transgenose (...) which might be transmitted to the descent" [40]. Five years later, it issued an "Announcement on Gene Therapy" stating that "somatic cell gene research aiming at treating monogenic or serious disease" should be authorised, but deliberate genetic alterations of the germ cells should be "absolutely prohibited". This position was reiterated two years later when the Committee was advised on the bioethics bills.

The regulatory French framework for gene therapy also implemented an ethical control of the research projects for clinical trials. This is, as we previously saw, the role of the CCPPRB.

This double – sanitary and ethical – control were the prerequisites for the development of gene therapy, by the French politicians. We may observe that gene therapy has, beyond these "technical" stakes, had an influence on the French traditional policy-style.

2.3. Reviewing the French traditional policy-style

Facing a new "public problem", the first political question was to know whether and, if so, how gene therapy needs to be regulated. For the ruling elite, a legal framework was the basic condition to develop somatic cell gene therapy, which would improve public health and place France among the leaders on the global scene of biomedical research.

The stake was to determine the contents of laws and the nature of judicial rules. Should special norms be created, specifically designed for a new issue, which impact on human beings and their descents have never been thought before? Or would it be sufficient to extend the traditional judicial regime, for example that of on human rights? Were specific penalties required? We previously saw that the choice made by the legislator was ambiguous since gene therapy products and clinical trials were regulated under existing regimes with major exceptions.

The main dilemma for the French legislator was to set limits without impinging upon the dynamics of science. This is, according to Hans Jonas, a new kind of responsibility for politicians. The specificity of "modern actions" implies, for Jonas, taking into account future times in the present decision. This is what the philosopher called a "constitutional state of transformation" [41]. In France, this had an impact on the traditional policy making. Trying to adapt to the dynamics of science required to set a flexible framework. To leave the door open to research, two new policy tools were experimented in France, breaking with the traditional features of policy making: a revisable legislative system and the regulation by an executive body.

The revisable legislative system was first introduced for the laws on abortion (in 1975, reviewed in 1979). It is significant that it was implemented for a highly controversial moral issue. The legislator, in 1994, decided that the bioethics laws should be reviewed within five years, after an assessment by the Parliamentary Office for Science and Technology (OPECST [42]), to adapt the judicial norms to scientific progress. By doing this, it tended to create a legislative dynamics. This is a very rare situation in France.

The regulation of gene therapy (in fact all the human health products) by an executive Agency, the Afssaps, is also a new feature in the French policy making, which is, by tradition, state-centred. The idea of setting an independent body, in charge of a specialised field, emerged in 1992, after the scandal of the contaminated blood, which had revealed that sanitary security could not be properly handled and managed at the ministerial level. An "Agency for Medicines" [43] was created, which scope was enlarged to all human health products in 1998, then replaced by the Afssaps. This kind of body is supposed to react in a faster and more appropriate way to scientific progress.

Gene therapy also raised questions on the decision-making process. Should the public be consulted on issues which would regulate their "essence"? In France, the decision-making process on the politics of science was, by tradition, the privilege of politicians, advised by scientific experts. We may observe that the recent developments of biomedicine disrupt this model. The 1988 law on the protection of people involved in biomedical research stipulated that the membership of the CCPPRB should include representatives of the "civil society" (here opposed to scientists). This openness towards society is indeed weak (the decree said that only 3 out of 12 members should be qualified in ethics, social field and law [44]) but it is a first step. It is more significant that in 2001, when the debate on transposing the European directive on clinical trials emerged, the Ministry of Health launched a public consultation on the Internet.

Identifying the political stakes gene therapy raised in France is leading us to broadening our scope and trying to sketch out some prospective paths.

3. BROADENING THE SCOPE: GENE THERAPY FROM A PROSPECTIVE POLITICAL POINT OF VIEW

After a decade of hopes and some disillusions raised from the therapeutic applications of gene therapy and the building of a legal framework for gene therapy, what could the forthcoming political stakes of gene therapy be? It seems unlikely that the current legislative framework for somatic cell gene therapy (products and clinical trials) be, at short term, modified. The announcement of the – relative – defeat of gene therapy in the clinical trial on "bubble babies" has not raised any doubt about its efficiency and its adequacy. There may be three paths for gene therapy as a forthcoming political stake: a regulation under a global binding ethical framework, the enlarging towards the authorisation of germ-line gene therapy and the implementation of a participatory democracy on biomedical issues.

3.1. Towards a global binding ethical framework?

Facing the growing trend of discussing biomedical issues on an international level, it is not mere fantasy to think that gene therapy might be regulated through a global binding ethical framework. International organisations, either global or supranational (e.g. the Council of Europe or the European Union), have already passed many recommendations on gene therapy, for example those banning techniques on germ-line cells while backing somatic cell gene therapy for research and therapeutic purposes. These were non-binding acts. In contrast, the convention on biomedicine elaborated by the Council of Europe in 1997 [45] was a binding act, when ratified by Member States. France has signed it but not ratified yet. The Convention stipulates that "an intervention on the human genome shall be authorised for preventive, diagnostic or therapeutic purposes and on condition that it would not modify the genome of the descent" (Art. 3).

Anticipating this trend, it is not unlikely that gene therapy would become a supranational political issue, over which member states would lose their sovereign right to decide upon the issue at hand. The United Nations are debating over implementing a universal bioethics convention which would not, however, be a binding act. The controversial debates on human cloning revealed the difficulty in building a consensus on such a topic.

National interests and differences in cultural backgrounds seem to prevail as obstacles to this cause. The problem lies in translating national perceptions into a common, global language. We may refer here to the three stages of inquiry [46] distinguished by Alfred North Whitehead: "romantic", "precision" and "generalising". At the "romantic" stage, "vague and hyperbolic arguments attract attention". This was the situation for gene therapy at the beginning of the 1980s, when the Human Genome Project was launched and that gene therapy was presented as a miracle way to fight against many physical (and even social) ills. At the "precision" stage, advantages and disadvantages of prospective technologies are examined with heightened reflection. This happened in France in the mid-1990s when gene therapy was set on the governmental agenda. Finally, at the "generalisation" stage, the issue would appear on a supra level. A prospective ethics, based on a global protection of the next generations, would be sketched out.

The globalisation of gene therapy would require sharing a common risk perception of gene therapy, especially the moral risk. We would reach, following the typology by R. Kielstein [47], the upper level on the scale of the acceptability of a new technology. This author identifies six levels: individual, familial, religious, societal, national and international [48]. Vaccinations against smallpox, measles, poliomyelitis and mumps have now been endorsed by most of the countries in the world but what would the situation be for gene therapy (even on the less controversial part of it, namely somatic cell gene therapy)? Harmonising cultural diversities on assessing moral risk seems a hard task to be handled. Mary Douglas and Aaron Wildavsky [49] pointed out that risk perception was embedded in cultural groups. Each group has its own perception of risk, as a way to identify in societies (this is what the authors called the "group" effect). Even inside each group, risk perception differs (the "grid" effect) according to its structure (either egalitarian or hierarchical).

Considering this on the international level, the number of groups is multiplied and there is no global government (or authority) to enforce the decisions, we may become sceptical about such a future for gene therapy.

3.2. On the societal side: Towards a social claim for authorising germ-line gene therapy?

Beyond somatic gene therapy, we may also anticipate a legal authorisation of germ-line cell therapy in France. In the field of biomedicine, several examples have shown that the strict limits fixed on a first step were later reviewed and enlarged, under the pressure of a societal claim.

The preimplantation genetic diagnosis (PGD) may relevantly illustrate this. Authorising the PGD was, in France, when it was put on the political agenda at the beginning of the 1990s, one of the most controversial issues in the public debate. Deputies and senators deliberated lengthily before agreeing on the issue. The National Assembly had authorised it before it was rejected by the Senate. A consensus was finally reached in 1993 when a new majority was formed in the National Assembly. PGD was authorised under strict circumstances, for therapeutic purposes (serious and incurable genetic inherited diseases) and on condition that it should benefit the sole embryo. Ten years after, in a consensual and serene atmosphere, Parliament made "saviour siblings" legal, a very different concept to PGD. This amendment was introduced on the parliamentary agenda after a doctor, Pr. René Frydman, asked the National Consultative Committee on Ethics (CCNE) in 2001 upon the morality of using a PGD for tissue-typing (in case of tissue typing, for a Fanconi disease). Facing this evolution of the legislative framework regulating PGD, it would not be surprising that germ-line cell therapy should be authorised in France. After all, the 2004 bioethics bills authorised the import of embryonic stem cells and the research on surplus *in vitro* embryos (through a derogation: for five years).

Supporting this assumption is the fact that germ-line gene therapy has not been explicitly mentioned and, thus, forbidden in the laws. Article 16-4 of the Civil Code may even be interpreted as implicitly authorising modification on the human genome of the descent for preventive or therapeutic purposes. There is, then, an open political space for germ-line gene therapy. From the analysis of the parliamentary debates on this issue, it appears that it has never been discussed in depth. As MP J.-F. Mattei wrote in a report on bioethics to Prime Minister Edouard Balladur in 1993 [50], germ-line gene therapy, at this early theoretical stage, was too unsure a technique to be regulated for that time. Germ-line gene therapy was thus not rejected on ethical grounds.

Should germ-line gene therapy be authorised, the legislative framework would have to be reviewed. The legislator would have to set limits to this technique, to prevent eugenic practice. This would raise the basic question: why should human beings be improved? Would the legislator set a list of indications for this technique, as it is the case on some controversial techniques (e.g. PGD)? In Germany, in 1987, a report on genetic engineering adopted by the Bundestag recommended that a detailed list of hereditary diseases for which somatic cell gene therapy would be allowed was a necessary prerequisite [51]. The Parliamentary Assembly of the Council of Europe had made, five years before, the same proposal.

The central goal would be that of social acceptability. The statement of former Prime Minister Lione Jospin, in 2000, on embryonic stem cells seemed to us very significant of the attitude of the French society (as seen in Section 2.1): "Should society and patients be deprived of therapeutic progress on philosophical, spiritual or religious grounds? I am convinvinced that for the French, basic values help framing scientific progress without hindering it" [52].

We may, at last, presume that scientific choices may become growingly debated in a more open and democratic way.

3.3. Towards a participatory democracy?

The past few years have seen the development of participatory tools in Europe such as consensus conferences or public consultations. They were introduced for environmental or technological controversial issues [53]. Should this trend continue? Even in such a state-centred country as France, a citizens' conference has been attempted (on GM food, in 1998).

The ethical paradigm is probably not enough to legitimise the political decision by the citizens. Since the end of the 1970s, distrust by citizens has increased towards experts and politicians. For many observers, people want to participate in the policy making. We are not so convinced of this. However, it is doubtless that citizens would like to be consulted and that they might feel reassured by this kind of procedure, through which the decision is negotiated on the "agora", in a public way.

The United Kingdom has experimented with several public consultations on "red" biotechnology, which are organised by the Human Fertilisation and Embryology Authority (HFEA). The main advantage of this political tool stands in the openness of the decision-making process. Consensus conferences were criticised in France for not being representative of the whole population. It seems to us that on issues involving human species, the "agora" should be completely open. Even if this may be an ideal, public consultations by the Internet seems to us, at least, a satisfying compromise. Moreover, parliaments should play a stronger role in science communication to citizens.

This would probably also enhance the social acceptability of risk on gene therapy. Having the feeling that they were associated in the decision-making process, the citizens may become less averse towards risk and policy-making, since the final choice would be perceived as not imposed by a technocratic-political elite but the outcome of a negotiation. Back to the Antique Greece?

Certainly not. However, it is obvious that politics is definitely the art of discourse.

4. CONCLUSION

Since the end of the 1980s, major attempts have been made by the French ruling authorities to encourage somatic cell gene therapy. To tackle an issue with both physical and moral consequences, French governments responded by implementing a new policy-style. It seems, however, that gene therapy has not reached its end as a political issue. We even feel that it is standing at a watershed. The scientific progress in genetics (knowledge and application) is creating social conditions for a forthcoming emergence of demand for germ-line gene therapy. This would be a much more sensitive issue than somatic cell gene therapy. Galileo or Prometheus? The choice may be by the citizens. Policy-style, on controversial technology, seems to be as important as the decision itself. This conclusion is, after all, nothing new: the Brooks report by the OECD 34 years ago suggests the same predictive: "there is nowadays a kind of passionate reaction towards science and technology (...) policies should take into account" [54].

REFERENCES AND NOTES

[1] "Deux malades atteints d'un cancer ont subi une thérapie génique", Le Monde, 31 January 1990.
[2] J.-F. Mattei, La vie en questions: pour une éthique biomédicale, Rapport au Premier ministre, La Documentation française, Paris, 1994, p. 105.
[3] For e.g. "Amandine", the first French "test-tube" baby was born eight years before, on 24 February 1982, at the Beclere Hospital (Clamart).
[4] 1996 and 2000, conducted by the European Commission (DG XII).
[5] S. de Cheveigne, D. Boy and J.-C. Galloux, *Les biotechnologies en débat. Pour une démocratie scientifique*, Balland, Paris, 2002, p. 124.
[6] Sofres/Ministry of Research, Feb. 2001, in Fournet, P., Les Français face aux nouveaux enjeux de la bioéthique, *La Revue française des sondages*, **179**(novembre) 2001, p. 8.
[7] Sofres/*L'Equipe*, 5/1/2001.
[8] Sofres/Ministry of Research, Feb. 2001, *in* Fournet, P., Les Français face aux nouveaux enjeux de la bioéthique, *La Revue française des sondages*, **179**(novembre), 2001, p. 6.
[9] P. Peretti-Watel, *Sociologie du risque*, Armand Colin, Paris, 2000, 286 p (coll. U).
[10] interview, 9 October 2002.
[11] "La thérapie génique au secours des artères bouchées", *L'Expansion*, 01/21/1999.
[12] Afssaps.
[13] Great Osmond Street Hospital.
[14] D. Kahneman and A. Tversky, Judgment under uncertainty: Heuristics and biases, in *Utility, Probability and Human Decision Making* (D. Wendt and C. Vlek, dir), Reidel Publishing, Dordrecht, 1975, pp. 141–162.
[15] P. Slovic, B. Fischhoff and S. Lichtenstein, Nous savons mal évaluer le risque, *Psychologie* (septembre), 1980, 45–48.

[16] Law "relative à la protection des personnes qui se prêtent à des recherches biomédicales" known as the "Huriet-Serusclat" law, n° 88–1138, 12/20/1988, modified in 1990, 1991, 1994 and 2001.
[17] Directive 2001/20/CE, 04/04/2001, European Parliament and Council; *JOCE* n° L121, 05/01/2001.
[18] By the investigator (or the doctor representing him or her) about the aim, the methodology and the duration of the research; the expected risk and benefit; the advise of the consultative committee. The consent may be withdrawn at any stage of the trial (Art. L. 1122-1 PHC). Penalty: three years imprisonment and a fine (Art. 223-8 Penal Code).
[19] Direction and control by a doctor with adequate qualification; in accordance to scientific methodology/code of practice; security of workers participating to the clinical trial.
[20] Law "relative au traitement des données nominatives ayant pour fin la recherche dans le domaine de la santé et modifiant la loi n°78-17 du 6 janvier 1978 relative à l'informatique, aux fichiers et aux libertés", n° 94-548, 07/01/1994.
[21] Law "relative au respect du corps humain", n° 94-653, 07/29/1994.
[22] Law "relative au don et à l'utilisation des éléments et produits du corps humain, à l'assistance médicale à la procréation et au diagnostic prénatal", n° 94-654, 07/29/1994.
[23] Title I Civil Code, "Du respect du corps humain".
[24] Title VI "Gene and Cell Therapy products and annex Therapeutic products".
[25] Art. L. 1261-1 Public Health Code.
[26] Art. L. 1211-1: concerns "the biological products with therapeutic effects, obtained from preparations of human or animal living cells".
[27] loi n° 96-452.
[28] "Haut Conseil des thérapies géniques et cellulaires".
[29] Law "relative à la bioéthique", n° 2004-800, 08/06/2004.
[30] Commission du Génie Génétique, placed under the authority of the Minister of Research, in charge of risk assessment (confine).
[31] Commission du Génie Biomoléculaire, placed under the authority of the Minister of Agriculture, to fix the duration of confine and the tests to be done for the patient to escape the confined place.
[32] The investigator is the "physical person(s) conducting the research"; the "promoter" is "the physical or moral person taking the initiative of the research" Art. L. 209-1, HPC.
[33] Respect to the basic conditions to preserve the patients' protection, information, consent, indemnities, relevance of the project, goals adequate with the expected results, investigator's adequate skills: Art. L. 209-12, al. 3, PHC.
[34] Penalty: 1 year imprisonment and 15,000 Euros fine – Art. L. 1126-5, PHC.
[35] H. Jonas, *Le principe responsabilité. Une éthique pour la civilisation technologique*, Cerf, Paris, 1997, p. 36. Translation by ourselves.
[36] L. Sfez, *Le rêve biotechnologique*, PUF, Paris, 2001, 127 p (coll. Que sais-je?).
[37] PHC, Art. L 152-8.
[38] Recommendation n° 934 on Genetic Engineering, 1982.
[39] Recommendation n° 1100 on Embryo Research, 1989.
[40] "Avis relatif aux recherches sur les embryons humains in vitro età leur utilisation à des fins médicales et scientifiques", 12/15/1986.
[41] p. 166.
[42] Office Parlementaire d'Evaluation des Choix Scientifiques et Techniques. This body was created in 1983, to advise Parliament (and the public opinion) on scientific issues. Its members are both deputies and senators.
[43] Agence du Médicament.
[44] Art. R. 1123-1 to R. 1123-23, PHC.
[45] Convention for the protection of human rights and dignity of human beings with regard to the application to biology and medicine, Oviedo, 04/04/1997.
[46] E. T. Juengst, Germ-line gene therapy: Back to basics, *J. Med. Polit.*, **16**(6), 1991, 587–592.

[47] R. Kielstein, Cultural and individual risk perception in human germ-line gene therapy, *Res., Polit.Life Sci.*, **13**(2), 241–243.
[48] Kielstein, R., *op.cit.*, p. 241.
[49] M. Douglas and A. Wildavsky, *Risk and Cutlture? An Essay on the Selection of the Technological and Environmental Dangers*, University of California Press, Berkeley, 1982, 221 p.
[50] J.-F. Mattei, *La vie en questions: pour une éthique biomédicale, Rapport au Premier ministre*, La Documentation française, Paris, 1994, p. 105 (coll des rapports officiels).
[51] Catenhusen, Wolf-Michael; Neumeister, Hanna (Hrsg.), Enquete-Kommission des Deutschen Bundestages, *Chancen und Risken der Gentechnologie. Dokumentation des Berichts an den Deutschen Bundestag*, München: Schweitzer Verlag, 1987, p. 183.
[52] *Le Monde*, 11/28/2000.
[53] S. Joss and S. Bellucci (dir.), *Participatory Technology Assessment. European Perspectives*, Centre for the Study of Democracy, London, 2002, 308 p.
[54] OCDE, *Science, croissance et société, une perspective nouvelle, Rapport du groupe spécial du Secrétaire général sur les nouveaux concepts des politiques de la science*, Paris, 1971, p. 15.

FURTHER READING

C. Bachelard-Jobard, *L'eugénisme, la science et le droit*, PUF, Paris, 2001, 341 p.
M. W. Bauer and G. Gaskell, *Biotechnology: The Making of a Global Controversy*, Cambridge University Press, Cambridge, 2002, 411 p.
U. Beck, *Die Risikogesellschaft. Auf dem Weg in eine andere Moderne*, Suhrkamp, Frankfurt Main, 1986, 391 p.
J.-R. Binet, *Droit et progrès scientifique. Science du droit, valeurs et biomédecine*, PUF, Paris, 2002, 298 p.
P. Birambeau, *Téléthon, le meilleur de nous-mêmes*, Balland, Paris, 2003, 375 p.
A. L. Bonnicksen, National and international approaches to human germ-line gene therapy, *Politics Life Sci.*, 1994, **13**(1), 39–49.
D. Boy, *Le progrès en procès*, Presses de la Renaissance, Paris, 1999, 264 p.
D. Boy, Les attitudes du public à l'égard de la science, in *L'Etat de l'opinion 2002* (eds. O. Duhamel and P. Mechet), Seuil, Paris, 2002, pp. 167–182.
M. Callon, P. Lascoumes and Y. Barthe, *Agir dans un monde incertain, Essai sur la démocratie participative*, Seuil, Paris, 2001, 357 p.
CNRS, Séminaire du Programme Risques Collectifs et Situations de Crise, *Sociologie des sciences, analyse des risques collectifs et des situations de crise. Point de vue de Bruno Latour*, CNRS, Actes de la première séance, Paris, 15/11/1994, 131 p.
Colloque De La Villette, *Le savant et le politique aujourd'hui. Un débat sur la responsabilité et la décision*, Albin Michel, Paris, 1996, 209 p (Idées).
Commission Des Communautes Europeennes, *Sciences du vivant et biotechnologie: une stratégie pour l'Europe, Rapport d'avancement et orientations pour l'avenir*, Communication de la Commission au Parlement européen, au Conseil et au Comité économique et social européen, COM(2003)96 final, 5/3/2003.
Commissariat General Du Plan, Ministere De L'ecologie Et Du Development Durable, Ministere De L'economie, Des Finances Et De L'industrie, La décision publique face aux risques, La Documentation Française, Paris, 2002, 167 p.
E. Dhonte-Isnard, *L'embryon humain* in vitro *et le droit*, L'Harmattan, Paris, 2004, 206 p (coll. Ethique médicale).

M. Douglas and A. Wildavsky, *Risk and Culture? An Essay on the Selection of the Technological and Environmental Dangers*, University of California Press, Berkeley, 1982, 221 p.

J.-P. Duprat, A la recherche d'une protection constitutionnelle du corps humain: la décision, n°94-343-344 D.C. du 27 juillet 1994, *Les Petites affiches*, **149**(décembre), 1994, 34–40.

P. Fournet, Les Français face aux nouveaux enjeux de la bioéthique, *La Revue française des sondages*, **179**(novembre), 2001, 3–11.

France. Conseil D'etat, *Les lois de bioéthique: cinq ans après*, La Documentation française, Paris, 1999, 337 p.

G. Gaskell and M. W. Bauer (éds.), *Biotechnology 1996–2000. The Years of Controversy*, Science Museum, London, 2001, 339 p.

L. Halman Y a-t-il un déclin moral? Enquête transnationale sur la moralité dans la société contemporaine, *Revue internationale des sciences socials*, **145**(septembre), 1995, 477–500.

C. Hood, H. Rothstein and R. Baldwin, *The Government of Risk – Understanding Risk Regulation Regimes*, Oxford University Press, Oxford, 2001, 217 p.

H. Jonas, *Le principe responsabilité. Une éthique pour la civilisation technologique*, Cerf, Paris, 1997, 336 p.

S. Joss and S. Bellucci (dir.), *Participatory Technology Assessment. European Perspectives*, Centre for the Study of Democracy, London, 2002, 308 p.

E. T. Juengst, Germ-line gene therapy: Back to basics, *J. Med. Polit.*, **16**(6), 1991, 587–592.

D. Kahneman and A. Tversky, Judgment under uncertainty: Heuristics and biases, in *Utility, Probability and Human Decision Making* (dir. D. Wendt and C. Vlek), Reidel Publishing, Dordrecht, 1975, pp. 141–162.

R. Kielstein, Cultural and individual risk perception in human germ-line gene therapy, *Res., Polit. Life Sci.*, **13**(2), 1994, 241–243.

J. W. Kingdon, *Agendas, Alternatives and Public Policies*, Harper & Collins, New York, 1984, 240 p.

P. Lascoumes, Normes juridiques et mise en oeuvre des politiques publiques, *L'année sociologique*, **40**, 1990, 42–71.

N. Lenoir, *Aux frontières de la vie. Une éthique biomédicale à la française*, La Documentation Française, Paris, 1991, 477 p (coll. Rapports officiels).

N. Lenoir and B. Mathieu, *Les normes internationales de la bioéthique*, PUF, Paris, 2002, 127 p (coll.Que sais-je ?).

G. Majone, From the positive to the regulatory state: Causes and consequences of changes in the mode of governance, *J. Pub. Policy*, **17**(2), 1997, 139–167.

G. Majone, The regulatory state and its legitimacy problem, *West Euro. Polit.*, **22**(1), 1999, 1–24.

J. G. March and J. P. Olsen, *Democratic Governance*, Free Press, New York, 1995, 293 p.

B. Mathieu, Force et faiblesse des droits fondamentaux comme instruments du droit de la bioéthique: le principe de dignité et les interventions sur le génome humain, *Revue du droit public*, **1**(janvier–février), 1999, 93–111.

J.-F. Mattei, *La vie en questions: pour une éthique biomédicale, Rapport au Premier ministre*, La Documentation française, Paris, 1994, 230 p (coll des Rapports officiels).

C. Z. Mooney (ed.), *The Public Clash of Private Values, The Politics of Morality Policy*, Chatham House Publishers, New York, 2001, 283 p.

P. Muller, L'analyse cognitive des politiques publiques: vers une sociologie politique de l'action publique, *Revue française de science politique*, **50**(2), 2000, 189–207.

D. Nelkin, *Controversy. Politics of Technical Decisions*, Sage, Beverly Hills, 1979, 256 p.

D. Nelkin and S. Lindee, *La mystique de l'ADN. Pourquoi sommes-nous fascinés par le gène?* Belin, Paris, 1998, 318 p (coll Débats).

D. Nelkin, *Technological Decisions and Democracy, European Experiments in Public Participation*, Sage, Beverly Hills, 1977, 110 p.

OCDE, *Science, croissance et société, une perspective nouvelle, Rapport du groupe spécial du Secrétaire général sur les nouveaux concepts des politiques de la science*, Paris, 1971, 125 p.
Office Parlementaire D'evaluation Des Choix Scientifiques Et Technologiques, *Le clonage, la thérapie cellulaire et l'utilisation thérapeutique des cellules embryonnaires*, Assemblée Nationale, Paris, Vol. 2198, 2000, 181 p.
J.-G. Padioleau, *L'Etat au concret*, PUF, Paris, 1982, 222 p.
J.-G. Padioleau, La société du risque, une chance pour la démocratie, *Le Débat*, **209**(mars–avril), 2000, 39–54.
P. Papon, Le temps des ruptures. *Aux origines culturelles et scientifiques du XXIè siècle*, Fayard, Paris, 2004, 329 p (coll. Le Temps des sciences).
A.-S. Paquez, La politique européenne des biotechnologies: les défis de l'évolution vers une communautarisation, in *L'Europe et les biotechnologies: urgences et impasses d'un débat démocratique* (dir. M. Aligisakis), Euryopa, Genève, 2004, pp. 161–202 (324 p).
G. Parayil, *Conceptualizing Technological Change. Theoretical and Empirical Explorations*, Rowman & Littlefield Publishers, Lanham, 1999, 210 p.
P. Peretti-Watel, *Sociologie du risque*, Armand Colin, Paris, 2000, 286 p (coll. U).
J. Plantier (dir.), *La démocratie à l'épreuve du changement technique*. Des enjeux pour l'éducation, L'Harmattan, Paris, 1996, 319 p (Logiques sociales).
P. Rabinow, La recherche génétique et la connaissance du vivant: un regard ethnographique sur le débat français, *Esprit*, **5**(mai), 2002, 132–144.
P. Rouvillois and G. Le Fur, La France face au défi des biotechnologies: quels enjeux pour l'avenir? *Journal Officiel de la République Française, Avis et Rapports du Conseil économique et social*, **13**(Part II), 1999, 1–284.
F. Salat-Baroux, *Les lois de bioéthique*, Dalloz, Paris, 1998, 199 p (coll. Dalloz Service).
L. Sfez, *Le rêve biotechnologique*, PUF, Paris, 2001, 127 p (coll. Que sais-je?).
P. Slovic, B. Fischhoff and S. Lichtenstein, Nous savons mal évaluer le risque, *Psychologie* (septembre), 1980, 45–48.
A. Somit and S. Peterson (eds.), *Human Nature and Public Policy, An Evolutionary Approach*, Palgrave Macmillan, New York, 2003, 266 p.
D. Stone, Causal stories and the formation of policy agendas, *Political Science Quaterly*, **104**(2), 1989, 281–300.
B. Sturlese, *Les sciences de la vie et le droit de la bioéthique*, ENM, Bordeaux, 1995, 307 p (coll. Essais et recherches judiciaires).
D. Tabuteau, *La sécurité sanitaire*, 2è éd, Berger-Levrault, Paris, 2002, 389 p.
M. Weber, *Le savant et le politique*, Paris, Vol. 10/18, 1963, 221 p.

EXTRACTS FROM INTERVIEWS

Dr. Marina Cavazzana-Calvo, scientist, investigator (with Pr. Alain Fischer) of the gene therapy clinical trial on "bubble babies", 9 October 2002 and 2 December 2003.

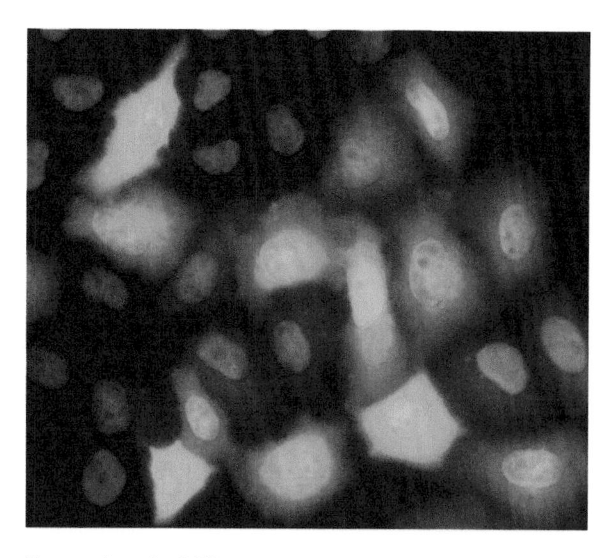

Expression of a GFP-tagged recombinant fusion protein and localisation of Hoechst-stained nuclei in human carcinoma cells

Section II
Ethical and Legal Aspects

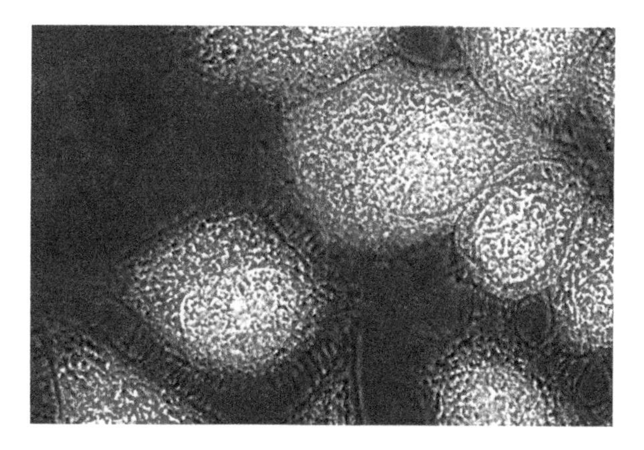

Fluorescence staining of endogenous STAT1 proteins in human cancer cells.

Section III
Perception and Communication

How to Communicate Risks in Gene Therapy?*

Andrea T. Thalmann

Swiss Federal Office for Public Health, Schwarztorstr. 96, CH-3003 Bern, Switzerland

Abstract

A clear and transparent risk communication is needed to reach an adequate understanding of gene therapy and of possible non-intended side effects and in order to provide a considerable basis for informed and responsible decision making. A main precondition for making responsible decisions is the adequate understanding of the risk or of the uncertainties regarding a new technology and of the existing evidence supporting or mitigating possible consequences on all relevant levels and among all affected parties/stakeholders. This chapter considers core problems of risk communication for the purpose of developing information strategies with regard to gene therapy. Examples are: the multidimensionality of laypersons' risk concepts, the role of heuristics for risk judgments, laypersons' perception of verbal uncertainty descriptions or the clarity of evidence characterization.

The general conclusion for risk communication on gene therapy refers to a more layperson-oriented approach with regard to risk information. Any risk information on gene therapy has to be carefully evaluated with respect to their clarity, and their appropriateness before being implemented.

Keywords: risk communication, risk perception, heuristics, emotionality, uncertainty, evidence characterization, clarity, comprehensibility, transparency.

Contents

1. Introduction 204
2. Definitions – clarity with regard to risk, hazards and uncertainty 205
3. Goal of risk communication 206
4. Core problems of risk communication 207
   4.1. Risk concept of experts and of laypersons 208
      4.1.1. Multidimensionality of risk 208
      4.1.2. Emotionality of risk information 208
      4.1.3. The role of heuristics for the interpretation of risk information 209
   4.2. Misunderstanding or misinterpretation of uncertainty 211
      4.2.1. Misinterpretation of numerical description of probabilities 211
      4.2.2. Laypersons' perception of verbal description of probabilities 213
      4.2.3. Laypersons' perception of verbal description of the strengths of evidence 214

---

* The views expressed in this article are entirely those of the author and may not in any circumstances be regarded as stating an official position of the Swiss Federal Office for Public Health.

4.3. The role of the communication of uncertainty in the scientific knowledge for risk perception 215
5. Conclusion for communication regarding gene therapy 216
References 217

1. INTRODUCTION

Gene therapy is an emerging technology characterized by uncertainty in scientific knowledge with regard to benefits and risks. The development of a routinely used gene therapy to successfully fight against severe diseases, such as cancer, takes longer than it had been expected during the first period of gene enthusiasm and might, in fact, turn out to have been "wishful thinking".

After an enthusiastic initial phase of clinical gene therapy studies, the achievements of gene therapy are perceived as rather disillusioning because of severe adverse effects shown in clinical trial (e.g. Gelsinger case in the US). As a consequence of these events, many of the clinical gene therapy tests are terminated due to different reasons such as the severe side/adverse health effects that have occurred, the media and public response which followed after the Gelsinger case, for example, or the lack of financial support as an answer to the changed risk perception of this new technology.

The technology is still in its early stages. Many questions regarding possible unintended (adverse) effects remain open and unresolved and lead to a serious criticism of the medical application and to controversial discussions. For example, concerning risk assessment, two questions are in the centre of the discussion. (a) The first question refers to the use of virus (gene taxis) to channel gene information into the cell, which is suspected to cause adverse consequences and hence pose risks. (b) The second question refers to the long-term consequences regarding a hereditary transmission of the gene modification. These questions are as yet unresolved. The scientific knowledge is incomplete and the medical technology is premature.

Decisions – among experts as well as laypersons and society – regarding the future of this promised medical innovation have to be made under conditions of uncertainty. A major precondition for making responsible and informed decisions – on both levels, experts and laypersons – is that an adequate understanding of the risk or of the uncertainty, respectively can be achieved.

A clear and transparent risk communication is needed to reach an adequate understanding of any innovation and of possible non-intended side effects and in order to provide a considerable basis for informed and responsible decision making.

How to reach a reliable basis for successful informed decision making for all people involved in the relevant risk assessment? Experience with other controversial discussions on new technologies and innovations in the past elucidate that it is not enough to give the right facts and numbers, even when they exist, to create a considerable basis for responsible decision making and risk assessment. Facts and figures have to be explained, taking in account experts' and laypersons' perspectives, thinking and concepts. The research on risk communication and risk perception gives important insight what should be considered while constructing risk-related information.

This paper will start by clarifying risk terminology and an introduction to the purpose of risk communication. Then specific aspects will be described which have to be taken into consideration in order to communicate about possible consequences of gene therapy.

2. DEFINITIONS – CLARITY WITH REGARD TO RISK, HAZARDS AND UNCERTAINTY

The *definition of risk* depends on the research field (e.g. technical risk analysis, cancer research, toxicology, epidemiology or law) in which it is used. Technical risk analysis refers to the hazard potential of industrial plants by particular materials, which causes health or financial damage by toxic or radioactive effects, etc. (see reference [1]). Epidemiology research describes risk as the relation between damages observed in one population compared to the one of another population (e.g. by relative risk or odds ratio).

A more research field independent conceptual definition of risk is used by the Scientific Steering committee's Working Group on Harmonisation of Risk Assessment Procedures of the European Commission: "The probability and severity of an adverse effect/event occurring to man or environment following exposure, under defined conditions, to a risk source(s)" [2]. In the following, the term "risk" will be used according to this definition. Risk is defined as a function of hazard and probability (see for an overview with regard to definitions of risk in the reference [3]).

Before going in detail, the meaning of the key concepts of hazard, risk and uncertainty should be defined more precisely: There is a *conceptual distinction between hazard and risk.* Contrary to daily language, where both concepts are used synonymously, scientific risk assessment distinguishes between the terms. *Hazard* refers to the ontologic potential of a risk source (e.g. technology, product, substance, activity, etc.) that might cause adverse effects.[1] *Uncertainty* refers to a basic characteristic of risk. We distinguish

[1] The WHO [76] defines "adverse effect" as change in morphology, physiology, growth, development or life span of an organism, which results in impairment of functional capacity or impairment of capacity to compensate for additional stress or increase in susceptibility to the harmful effects of other environmental influences. Decisions on whether or not any effect is adverse require expert judgement.

between different types of uncertainty. Uncertainty that refers to incomplete knowledge about a phenomenon, process or measurement is defined as epistemic *uncertainty* and has to be separated from variability, which describes the aleatoric uncertainty. In principle, epistemic uncertainty, on the objective's or empirical side, can be reduced through further research. Variability refers to the heterogeneity within a population (the true/real diversity within a population). As factually given, variability cannot be reduced through further research (cf. taxonomy of uncertainty by NRC [4,5]).

Another distinction should be made between uncertainty, vagueness and ambiguity. A matter is vague if it is not clearly defined or not capable of being understood precisely. Vagueness describes an uncertainty, which is not precisely known. In this case there is uncertainty about the uncertainty. Vagueness has to be separate from ambiguity. An event is ambiguous if it can be understood in two or more different, but precise ways [6].[2]

According to the classification of Wiedemann *et al.* [7],[3] we have to deal with an *unclear risk* if scientific knowledge is incomplete with regard to the cause-and-effect relationship between exposure and damage. Regarding an unclear risk, for example, it is unclear whether the risk really exists. This uncertainty can be traced back to the lack of scientific evidence that would positively corroborate any adverse effect. If no adverse effects can be described, however, there is no justified reference to risk.

Considering the risk assessment of new technologies, this problem is one of the key issues. It pertains to the general question, whether risks (and if any, with what kind of hazard potential) are derived from new technologies, and how to assess them. One main point is here, that the related scientific knowledge is incomplete with regard to its potential adverse consequences of new technologies. Therefore, conclusions about and implications for consequences for human beings and the environment must remain preliminary. Dealing with such uncertainties in knowledge is especially not only difficult for laypersons, but for experts as well. Current discussions about innovations – as, e.g. gene therapy or already existing technology controversies such as in the debate on mobile telecommunication technology – are pertinent examples in this dilemma.

3. GOAL OF RISK COMMUNICATION

Risk communication is defined as the "process of exchange about how best to assess and manage risks among academics, regulatory, practitioners, interest groups and the general public" [8].

---

[2] A clear terminology between vagueness and ambiguous is necessary. Unfortunately, the terminology of both is often used wrongly or vice versa in decision theory or uncertainty research.

[3] See classification of Wiedemann *et al.* [7], who differs between three types of risks: (a) known risks are risks for which the cause-and-effect relationship between exposure and damage is proven, (b) unclear risks are risks for which a cause-and-effect relationship between exposure and damage is assumed, but not yet sufficiently substantiated, and (c) unknown risks refers to a total lack of knowledge about new risks.

This definition shows distinctly the bi-directional character of risk communication. It is not a one-way communication from expert to layperson (cf. [9]). Moreover, risk communication is an "interactive process of exchange of information and opinion among individuals, groups and institutions. It involves multiple messages about the nature of risk and other messages, not strictly about risks, that express concerns, opinions or reactions to risk messages or to legal and institutional arrangements for risk management"[4] [10].

Risk communication describes an interactive exchange between different participants dealing with critical points in the whole risk and innovation management. It is a network of mutual and respectful relationship among individuals in a risk management process (cf. [11]).

Ways of risk communication range from simple communication of risk information through different steps of sophistication to the implementation of whole participatory approaches. The chief goal is that all participants of a risk management process understand the relevant information and are empowered to make their own decision regarding particular risk issues.

Based on a respectful relationship, risk communication promotes the understanding of risks and related facts, and by promoting this risk communication it provides the preconditions for informed and responsible decisions and might lead to changes of behaviour and attitudes.

Multiple recommendations exist with regard to what and how to communicate on new technology and innovations as e.g. gene therapy. A key point to consider is the fact that risk communication should not only be based on personal and daily experience, but it also has to be "evidence based". In other words, the impact of risk communication should be evaluated before implementing it in practice. This elucidates from the fact that "ill" risk communication not only fails, but also often produces further costs and negative side effects (cf. [12]).

4. CORE PROBLEMS OF RISK COMMUNICATION

There is no unique agreement on how to communicate effectively (cf. [13]). Two major characteristics of a successful risk communication are comprehensibility and transparency (cf. [14]).

Theory and practice of the risk research field identified several core problems of risk communication, which have to be considered to achieve transparent and clear communication. One refers to the discrepancy between experts' and laypersons' concept of risk. Another is the misunderstanding or misinterpretation of uncertainty. Both will be discussed in the following.

---

[4] Risk management is the process of identifying, evaluating, selecting and implementing actions to reduce risk to human health and to ecosystems. The goal of risk management is scientifically sound, cost-effective, integrated solutions that reduce or prevent risks while taking into account social, cultural, ethical, political and legal consideration (The Presidental/Congressional Commission on Risk Assessment and Risk Management, 1997).

Further complementary points to be considered in risk communication are the role of trust and the role of social amplification. Both influence the effectiveness of risk communication crucially. A good overview regarding trust is given by Cvetkovich and Löfstedt [15]. Regarding social amplification and the role of media, see the overview given by Kasperson *et al.* [16] and by Pidgeon *et al.* [17].

4.1. Risk concept of experts and of laypersons

A core problem for risk communication refers to the discrepancy in risk concept between experts and laypersons. Experts and laypersons perceive risks differently. The reasons for this discrepancy are multiple. Wrong understanding and interpretation of risk knowledge, and a lack of risk knowledge are certainly important explanations. Another reason (and the major one) is the different approaches to assess risks.

4.1.1. Multidimensionality of risk

Experts base their risk assessment on objective parameters such as probability, hazard potential, possible chains of cause-and-effect. Under an expert perspective, questions such as the following have to be raised to assess risks: Is there a potential hazard and what is its nature? What dose will induce a harmful effect? Who is exposed to what dosage? How significant is the risk? (see reference [18]).

In contrast to this technical-scientific risk assessment, laypersons' approach is characterized by an intuitive and holistic way to assess risk. Laypersons want to know yes, there is a danger or no, there is nothing. Laypersons are not interested in average, they are personally shocked and evaluate risks under this personal perspective [8,19]. Several studies deal with the laypersons' multidimensionality of risk concept and explain it by different theoretical conceptual approaches [77].

While probability and hazard are the parameters to assess risks under scientific tradition, laypersons use further criteria to evaluate risks. For example, controllability, frightfulness, benefit or voluntary are important parameters, which influence laypersons' risk perception in a considerable way.

4.1.2. Emotionality of risk information

Laypersons' risk perception is based on their interpretation of events or of stories. In this connection, emotionality plays a critical role for laypersons' risk assessment. Risk description introduces emotionality. Risk stories are heavily influenced by the media by implementing risks in scandal stories, tragedies or disaster reports. The presentation of information causes specific emotional states in the recipients and act as a responsible parameter in risk perception. The findings of an experimental study of

Wiedemann and colleagues [18] show this distinctly by using the risk story approach. Confronting laypersons with identical risk stories (same type of risk and severity of consequences) except that one story arouse outrage and the other one leniency lead to different risk appraisal. A risk story arousing outrage leads to increased risk perception compared to the leniency condition. Although the "objective" risk information given in the stories was identical, the risks were judged rather differently. The context in which the story was implemented (story frame) influenced the risk judgement [18].

The role of emotionality for risk perception is shown by different studies. For example, the presentation of strongly negative consequences such as terrifying diseases are perceived as fear appeal that provokes negative emotions and influences information processing and risk judgements [27–29]. Highly emotional information – e.g. risk information about cancer – polarizes already existing beliefs and opinions [78]. Highly emotional information amplifies the risk appraisal of concerned people (based on their beliefs) and diminishes the risk appraisal of unconcerned people (based on their beliefs).

4.1.3 The role of heuristics for the interpretation of risk information

Laypersons' risk judgements are based on intuition and therefore sensible for fallibilities. If information is difficult to process people adopt mental "strategies" that simplify the task of judgements and of choice, then these strategies are often effective, but lead sometimes to systematic bias as well. The following examples attempt to visualize such systematic bias. An informative overview is given by Hogarth [30] or by Jungermann *et al.* [31].

Laypersons use sometimes a kind of "*unrealistic optimism*" to judge a risk. In this case, the subject assumes that a negative event or adverse effect could never happen personally but could reach others. This unrealistic optimism plays a crucial role for the assessment of the personal probability, e.g. regarding an infection disease (see study of Weinstein [32] regarding perceived risk of HIV infections). Due to a wishful thinking, the probability of pleasant events is overestimated and of unpleasant events is underestimated.

Another phenomenon "representativeness" refers to the heuristic that the probability for an event is higher if the event is very representative for a specific population or group. In other words, we tend to imagine that what we see or will see is typical of what can occur. Unusual events are not expected. It is postulated (see reference [33]) that new information will be compared to already existing information with regard to typical characteristics of an event or a group. The result of this comparison influences then the judgements. For example, imagining that someone is a lawyer because he exhibits characteristics typically of a lawyer [30]. This heuristics works in most of the cases, but leads to several biases. One example is the *misinterpretation of coincidence* characteristics. There is an expectation that characteristics of an infinite coincidence process will occur as well in very short finite sequences. Confronted with two possibilities of sequences regarding the birth of six children, most of the people assess the sequence "boy/girl/girl/boy/girl/boy" as more probable than the sequence "boy/boy/boy/girl/girl/girl". The second sequence looks not enough coincident because of their regularity, and therefore is less probable. People forget that both sequences are equally probable, because the events are independent from each other [33].

Such misinterpretation of coincidence plays perhaps a role for the perception of cluster regarding diseases or events. For example, people do not believe that an increase of a particular disease (e.g. leukaemia) near an industrial plant could be the result of a coincidence. But the observed link between plant and people suffering from the disease is not sufficient for a sound argument. Data with regard to the following aspects have to be included: the neighbours who did not fall ill, people suffering from the disease, but who did not live near the industrial plant and comparable people who did not have the disease and did not live near the industrial plant.

Biased judgements occur as well regarding the retrospective interpretation with regard to an outcome of an event. The knowledge regarding the outcome of an event may restructure memory in a way that people are not surprised about what had happened and they can easily find plausible explanations [30]. This *hindsight bias* describes the diminished ability to imagine alternative explanation for a past event. In several experimental studies this effect was shown. For example, subjects overestimate the amount of their own knowledge and expectation before a particular event if asked after the event [79].

The *availability* of risk information influences risk judgement and risk perception. The frequency of well-published events is overestimated (e.g. death due to homicide, cancer) and the frequency of less well-published events is underestimated (e.g. death due to asthma or diabetes). The degree of the subjective probability of an event or consequence depends on the easiness or rapidity of recall from the memory [30].

Further, people have their own representations regarding the cause– effect correlation or of the risk, respectively. Such *mental models*, peoples' subjective representations of how things proceed and work, are sometimes opposite to the real facts regarding risks and regarding how they will take place. Such biased mental models lead to wrong risk judgements [34].

The list of heuristic phenomena is quite long. While judging a technology, a medical therapy, an event or something else, people use very different heuristics to make a decision which is sometimes biased. Further examples of heuristics are overconfidence of the truth of their own answer, conjunction fallacy regarding causality, the illusion of control and so forth. A good overview is given by Hogarth [30].

4.2. Misunderstanding or misinterpretation of uncertainty

The communication of uncertainty[5] implies several problems with regard to how to present and how to describe uncertainty in a layperson-oriented and -comprehensible way. A considerable amount of literature and empirical studies exist regarding numerical and verbal description of probability and frequency, which evidences the problems and biases appearing when interpreting and referring to verbal and numerical uncertainty description. Neither numerical nor verbal description of probability or frequency offers an optimum of transparency and clarity. Moreover, there are only a few studies referring to the description and interpretation of "vague uncertainty" and of the strength of evidence, which implies additional problems for effective communication – in particular for risk communication of new technologies.

4.2.1. Misinterpretation of numerical description of probabilities

The lack of clarity of risk information is partially founded by a wrong or an inadequate choice of uncertainty descriptions. Not all numerical uncertainty descriptions are equally comprehensible. Some descriptions turn out to be "difficult" with regard to their correct interpretation by layperson due to a lack of numerical or statistical understanding. The descriptions by percentage units, which are very often used in media, are often misunderstood. A good example is the meaning of "40%": In a study of Gigerenzer [35], 1000 people were asked about the meaning of 40%. Around one-third of the participants did not understand the percentage correctly. The comprehensibility of uncertainty description depends on the used risk unit (e.g. probability unit, frequency unit, relative risk) and plays a crucial role for assessing risks and the perception of risk [35,36].

Small probabilities, for example, lead to misinterpretation and are difficult to understand correctly [35,37–39]. Very small probabilities are overestimated and very high probabilities are underestimated (cf. [40]). Events or consequences with small probabilities demand high cognitive skills for adequate processing. This is one reason why people have difficulties to weigh small probabilities with potential consequences correctly [41].

One explanation for the lack of comprehensibility refers to the assumption that probabilities do not correspond to our thinking, and therefore are misunderstood [35]. Natural frequencies correspond more to peoples' thinking. A study investigating the understanding of the probability of breast cancer in women shows evidence for this hypothesis [42,43]. Only two of 24 physicians calculated the correct probability of illness based on probability

[5] See the definition of uncertainty, vagueness or ambiguity above.

description. In contrast, given the same data by natural frequencies most of the participants calculate the probability correctly or quite correctly. An improvement of the comprehensibility is reached by presenting probabilities in the form of natural frequencies. On the cognitive level, it is easier to understand, if one of 10,000 subjects, who took a medicament, develops side effects compared to an equal description in the form of percentages (the subject develops side effects with probability of 0.0001) (cf. [35]).

A risk can be described by the *relative risk* (RR), which is defined as the probability of an adverse effect or a disease/disorder of subjects who are exposed to a risk factor compared to subjects without risk exposition. The relative risk presents a unit for a relative effect of a certain exposition and it indicates the increased or decreased frequency of a certain disease of exposed subjects compared to unexposed subjects.

Risk description by relative risk influences the interpretation of the amount of benefit or damage compared to other risk units such as the incidence[6] rate, for example. The relative risk description lets numerical value appear bigger (regarding benefit or damage) compared to the identical numerical value described by the incidence [35]. This has an impact on the risk perception and risk behaviour. The description of risk information by relative risk increases risk-mitigating behaviour compared to the description of the identical risk information by the incidence rate (cf. [44]). The different impacts of the RR and of the incidence rate are more visible regarding extremely small values. Extremely small numerical values in the form of incidence rate are interpreted as near zero. The presentation of the identical risk entity by relative risk causes the opposite effect. The probability of a consequence is interpreted differently from zero. This has an impact on risk perception and might influence risk behaviour.

The neglect of the *a priori* base rate of an event leads to biased interpretation of a risk too. This biased judgement is known under *base-rate error.* People tend to ignore the base rate intuitively and use more case-specific information while interpreting the probability of an event. For example, people neglect the *a priori* probability of an event while estimating the probability of occurrence [30]. A tourist in a winter resort could be so impressed by the high amount of snow at 1000 m above sea level, that he forgets the general frequency of snowing in April. Therefore, his subjective estimation of the probability of snowing for the next day will be higher compared to the estimation based on the true base rate. The base rate helps to interpret numerical risk units more correctly and serves as a frame to interpret risk entities [45]. Further, the knowledge about the base rate plays a role with regard to the interpretation of verbal uncertainty description (see next chapter regarding verbal description).

[6] Incidence is defined as the frequency of occurrence of a disease within a certain time period, independent if the disease still exists at the end of the time period.

4.2.2. *Laypersons' perception of verbal description of probabilities*

The use of verbal description of probabilities and frequencies leads to a lack of clarity and transparency of risk information. Several studies present evidence for the ambiguity of verbal probability description [46–51] and for the different impact of verbal description on risk judgements compared to numerical ones [52].

Verbal descriptions of probability such as "rarely", "sometimes" and "likely" are interpreted differently by laypeople. Verbal descriptions of probability are equated with other numerical values by layperson compared with the intended numerical meaning by the information source [48,53]. The *ambiguity of verbal probability and frequency description* is the object of several studies (cf. [46,48,54–58]). The subjective meaning of verbal probability or frequency expression such as probable, frequent and possible often varies from person to person.

Asking people to equate numerical value for "unlikely" on a scale from 0.00 to 1.00, the result will be a wide range from 0.09 to 0.30, which is quite similar to the estimation for "somewhat likely" (from 0.09 to 0.45). Each subject will equate another numerical value and will interpret the meaning differently (cf. [59]). The variability in the interpretation of verbal probability descriptions refers not only to laypersons but also to experts. This is nicely shown by the study of Tavana *et al.* [60]. Finance experts had to equate verbal expression such as "fair chance" or "unlikely". The received numerical value spread very widely from 0.18 to 0.66 for "fair chance" and from 0.05 to 0.45 for "unlikely". Such big inter-individual variability regarding the interpretation of verbal description of probability or frequency provides problems for effective risk communication. The intended meaning of verbal descriptions cannot be controlled, and therefore the impact of risk communication cannot be controlled. There is evidence for the different significance of verbal descriptions for risk judgements, as compared to numerical descriptions [50,61].

Verbal descriptions of probability or frequency are often assessed as too "vague" [54,57,62–64]. Although there is no agreement with regard to the effectiveness of verbal and numerical description of probability or frequency [6,64], people prefer to receive numerical data, and prefer to communicate identical information by verbal descriptions. Different studies document this phenomenon under "*communication mode preference paradox*" [65,66]. Verbal descriptions of probability are associated with more vagueness and less precision regarding the uncertainty of the consequences when compared to numerical description [50]. Numerical descriptions of probability express more trustworthiness. For example, numerical frequency descriptions of side effects of a medical drug increase the willingness to choose a particular medical drug compared to a verbal uncertainty description of side effects, which decrease the willingness [48,53]. Other research shows evidence that the degree of perceived precision/vagueness influences the appraisal of the credibility of information [67]. With regard to "coarser" numerical description of uncertainty (by intervals) similar results are found [68].

The meaning of verbal frequency or probability descriptions depends highly on the *context* in which they are embedded. The numerical equivalent of verbal expression of frequencies such as "rarely", "sometimes" or "frequent" that are embedded in the patient package inserts of a medical drug differ from a neutral context [48]. The following example shows the context dependence very nicely. The numerical interpretation of "frequent" is totally different in the following two contexts: "going to the movie theatre frequently" and "visiting Asia frequently".

The meaning of verbal probability or frequency description is changed by contextual characteristics. For example, the knowledge of the base rate of an event (see above) influences the interpretation of verbal probability description [48,53,58,69]. Other characteristics of an event or of the context influence the interpretation of probability descriptions as well. A good example is the severity of an event or a consequence. Verbal probability descriptions of severe events are interpreted numerically higher compared to verbal probability description of less severe events [58]. The verbal frequency descriptions of severe side effects of a medical drug are interpreted higher compared to same frequency descriptions regarding slight side effects [48].

4.2.3. Laypersons' perception of verbal description of the strengths of evidence

Incomplete scientific knowledge poses several problems for risk communication. Due to the fact that it is not possible to define the uncertainty quantitatively (because of the incomplete knowledge), another decisive criterion for risk characterization has to be used. The strength of evidence for the existence, the absence of an adverse effect can be used to describe new technologies and their possible consequences. It has to be communicated how certain (or uncertain) the existing knowledge (or existing arguments support/mitigate the hazard potential) is.

Current *evidence characterizations* attempt to describe different strengths of evidence by verbal labels such as "suspicion", "consistent hint" or "weak hint". There is empirical evidence that such verbal evidence descriptions are highly ambiguous and vary from person to person. The study of Thalmann [78] tested evidence descriptions of current risk assessment and risk communications. Asking laypersons to equate different verbal evidence descriptions to numerical values on a scale from 0% to 100% of strength of evidence lead to discouraging findings: The subjects' estimations show wide spread for each verbal evidence description and overlap with other description of the strength of evidence. For example, the numerical estimation of the verbal evidence expression "consistent hint" spreads between 6% and 100% strength of evidence and is almost identical with the numerical estimations of the verbal evidence description "strong hint" (between 16% and 90%). Further, the numerical estimations of the subjects indicate that some verbal evidence descriptions were understood totally in the opposite way. Against the intention of the information source the verbal evidence description "suspicion" was equated to smaller numerical values by laypersons than the verbal evidence description "hint". In some limited extent, the use of additional definitions (explaining the meaning of the verbal labels of different strength of evidence) increased the clarity of the verbal evidence description. But some degree of ambiguity and variability remains. Verbal descriptions of evidence have an ambiguous meaning. Therefore, the use of verbal description of different strength of evidence is problematic and highly questionable for risk communication. More knowledge has to be gained on how to characterize and to communicate about different strength of evidence with regard to uncertain risks.

4.3. The role of the communication of uncertainty in the scientific knowledge for risk perception

The communication of uncertainty regarding new technologies or innovations includes two important questions: (a) On the level of strategies and tools the question is raised how to describe vague uncertainty. In other words, how to characterize different strength of evidence clearly and transparently. Current (verbal) evidence characterizations are insufficient. The problems that appear here are described in the chapter above and are still unresolved. (b) The other question is whether the communication of uncertainty in the scientific knowledge influences laypersons' risk perception (for example, amplify the perceived threat). Also this question is still open and refers to permanent and controversial discussion.

Although the current opinion demands for a clear communication of uncertainty and disagreements regarding risks and adverse consequences because of the moral duty to the public and because of the ideal of risk communication [23,70–72], there are relatively few and inconsistent empirical data with regard to the impact of uncertainty communication [73].

Some studies show evidence for a positive effect of the communication of uncertainty [52,61]. Governmental institutions that communicate uncertainty are assessed as more trustworthy by laypersons. Further, from the part of layperson there is a preference for clear information in respect to what is known and what is unknown. But the majority of the findings from the studies by Johnson and Slovic [80,81] and Johnson [73] present opposite evidence: the communication of uncertainty increases risk perception and distrust. And other studies regarding the impact of the disclosure of uncertainty in risk assessment show no effect on risk perception or trust in health protection (e.g. [74,75]).

Nevertheless, a final conclusion with regard to the impact of communication of the uncertainty in scientific knowledge is not possible based on the current existing scientific state of art.

5. CONCLUSION FOR COMMUNICATION REGARDING GENE THERAPY

The discussion on gene therapy involves experts from different scientific disciplines as well as laypersons from different background, worldviews or beliefs. Risk communication supports and promotes the process of assessing risk which should be scientifically sound and should describe integrated cost-effective solutions that aim to reduce or prevent risks while accounting for the relevant social, cultural, ethical, political and legal considerations.

Decisions – among experts as well as laypersons and society – regarding the future of this promising medical innovation have to be made under conditions of uncertainty. A main precondition for making responsible decisions is the adequate understanding of the risk or of the uncertainties regarding a new technology and their possible consequences on all relevant levels and among all affected parties/stakeholders. A clear and transparent definition and weighing of the existing evidence supporting or mitigating possible consequences such as health effects is necessary for assessing a risk. Risk communication should help to create such an adequate understanding and a considerable basis for informed and responsible decision making. Risk information should be clear and transparent, avoiding ambiguity and possible misinterpretation. These are the major conditions to reach a considered and an effective communication.

This chapter considers core problems of risk communication for the purpose of developing information strategies with regard to gene therapy. Some of the shown points lead to concrete solutions. For example, probability description by natural frequencies best accommodates the people's common ways of thinking, and is therefore more comprehensible. Other problems of risk communication – such as a clear and transparent evidence characterization – indicate the need for further research in the field of risk communication. And again other problems (e.g. laypersons' use of intuitive criteria to assess risk) cannot be changed, but indicate the need for developing better strategies to deal with. From this point of view, the following insights have to be highlighted with regard to gene therapy.

The multidimensionality of laypersons' risk concept cannot be changed in favour of an objective technical risk concept in risk communication. In fact, laypersons' risk concepts include important criteria, which have to be taken into consideration so as to reach an appropriately comprehensive discussion of gene therapy. The awareness of different risk concepts and of the multidimensionality of laypersons' risk concept in particular compared to those of experts' reflects the first step to create more common-sense-oriented information. Risk information should respond to the level and mode of comprehension on the side of addressees regarding knowledge, perception, interpretation and concepts of risk.

Emotionality presents one major impact factor of risk perception. Particular societal debates on new technologies are often characterized by a high level of emotionality. Gene therapy is an exemplary area for an emotional discussion. Particularly here the development of a balanced approach to risk information is needed, which would not give rise to fears but which would rather highlight scientifically sound and reasonable pro and contra arguments. That implies not to develop information, which denies existing emotions and feelings in the discussion about gene therapy. Balanced information about gene therapy should rather establish a constructive counterpart to an emotionally filled arena of the debate, which is very often created by the media hype with catching stories and headlines.

People use heuristics when making decisions, regarding gene therapy, too. These strategies are often effective, but sometimes lead to systematic bias. Accurate understanding of the facts, of the decision task and of the relation between important variables is needed to promote less biased and more scientifically sound decisions. A clear and transparent risk communication on gene therapy could contribute an important part to reach this, together with obvious implications for the related ethics debates.

One step to reach clarity and transparency in risk communication on gene therapy will be the improvement of adequate communication tools and strategies. An example is the development of better characterization tools for different strength of evidence. It is of major interest that all participants of a risk assessment process understand the existing uncertainties regarding new technologies and their possible consequences.

Another step will be the creation of sensitivity to the problem of biased decision making and misinterpretation of risk information among all participants of the risk-assessment process. A more objective and scientifically sound discussion on gene therapy will be promoted by the reflection of one's own biased decision making.

The general conclusion for risk communication and risk assessment on gene therapy refers to a more layperson-oriented approach with regard to risk information. Any risk information on gene therapy has to be carefully evaluated with regard to an addressee-oriented perspective before being implemented in technical reports, brochures or any other risk communication material, and used in policy making or for discussions.

REFERENCES

[1] U. Hauptmanns, M. Herttrich and W. Werner, *Technische Risiken: Ermittlung und Beurteilung*, Springer, Berlin, 1987.

[2] EC, First Report on the Harmonisation of Risk Assessment Procedures – Part 2 (report): Brussels: European Commission, Scientific Committee's Working Group, Directorate C, 2000.
[3] H. Schütz, P. Wiedemann, W. Hennings, J. Mertens and M. Clauber, *Vergleichende Risikobewertung. Konzepte, Probleme und Anwendungsmöglichkeiten*, Forschungszentrum Jülich, Jülich, 2004.
[4] NRC: National Research Council, *Science and Judgment in Risk Assessment*, National Academic Press, Washington, DC, 1994.
[5] M. G. Morgan and M. Henrion, Uncertainty, *A Guide to Dealing with Uncertainty in Quantitative Risk and Policy Research*, Cambridge University Press, Cambridge, 1990.
[6] D. V. Budescu, S. Weinberg and T. S. Wallsten, Decisions based on numerically and verbally expressed uncertainties, *J. Exp. Psychol.: Hum. Percept. Perform.*, 1988, **14**(2), 281–294.
[7] P. M. Wiedemann, C. Karger and M. Clauberg, *Machtbarkeitsstudie Risikofrüherkennung im Bereich Umwelt und Gesundheit*, Teilvorhaben 9: Aktionsprogramm Umwelt und Gesundheit" (im Auftrag des Umweltbundesamtes Berlin.) Forschungszentrum Jülich, MUT. Umweltforschungsplan des Bundesministeriums für Umwelt, Naturschutz und Reaktorsicherheit, F+E-Vorhaben, 200 61 218/09, 2002.
[8] D. Powell and W. Leiss, *Mad Cows and Mother's Milk: Case Studies in Risk Communication*, McGill-Queen's University Press, Montreal, 1997.
[9] H. Schütz and P. M. Wiedemann, Risikokommunikation als Aufklärung, Z.f. *Gesundheitswissenschaften*, 1997, **3**(Beiheft), 67–76.
[10] NRC: National Research Council, *Improving Risk Communication*, National Academy Press, Washington, DC, 1989.
[11] B. Fischhoff, Risk perception and communication unplugged: Twenty years of process, *Risk Anal.*, 1995, **15**, 137–145.
[12] B. Fischhoff, A. Bostrom and M. J. Quadrel, Risk perceptions and communication, *Ann. Rev. Pub. Health*, 1993, **14**, 183–203.
[13] A. J. Rothman and M. T. Kiviniemi, Treating people with information: An analysis and review of approaches to communicating health risk information, *J. Natl. Cancer Inst. Monogr.*, 1999, **25**, 44–51.
[14] K. C. Calman, The language of risk: A question of trust, *Transfusion*, 2001, **41**, 1326–1328.
[15] G. Cvetkovich and R. E. Loefstedt (eds.), *Social Trust and the Management of Risk*, Earthscan, London, 1999.
[16] R. E. Kasperson, O. Renn, P. Slovic, H. S. Brown, J. Emel, R. Globle, J. X. Kasperson and S. Ratick, The social amplification of risk: A conceptual framework, *Risk Anal.*, 1988, **8**, 177–187.
[17] N. Pidgeon, R. E. Kasperson and P. Slovic, *The Social Amplification of Risk*, Cambridge University Press, New York, 2003.
[18] P. M. Wiedemann, M. Clauberg and H. Schütz, Understanding amplification of complex risk issues: The risk story model applied to the EMF case, in *The Social Amplification of Risk* (eds. N. Pidgeon, R. E. Kasperson and P. Slovic), Cambridge University Press, New York, 2003, pp. 286–301.
[19] P. M. Wiedemann, EMF risk communication: Themes, challenges, and potential remedies, in *EMF Risk Perception and Communication* (eds. M. H. Rapacholi and A. M. Muc), WHO, Geneva, 1999, pp. 77–107.
[20] P. Slovic, B. Fischhoff and S. Lichtenstein, The psychometric study of risk perceptions, in *Risk Evaluation and Management* (eds. V. T. Covello, J. Menkes and J. Mumpower), Plenum, New York, 1986, pp. 3–24.
[21] A. Bostrom, B. Fischhoff and M. G. Morgan, Characterizing mental models of hazardous process: A methodology and an application to radon, *J. Soc. Issues*, 1992, **48**, 85–100.
[22] H. Jungermann, H. Schütz and M. Thuering, Mental models in risk assessment: Informing people about drugs, *Risk Anal.*, 1988, **8**(1), 147–155.

[23] D. G. MacGregor, P. Slovic and M. G. Morgan, Perception of risk from electromagnetic fields: A psychometric evaluation of a risk-communication approach, *Risk Anal.*, 1994, **14**(5), 815–828.

[24] M. G. Morgan, B. Fischhoff, A. Bostrom and C. J. Atman, *Risk Communication: The Mental Models Approach*, Cambridge University Press, New York, 2002.

[25] P. M. Wiedemann and H. Schütz, *Developing Dialogue-Based Communication Programs*, Arbeiten zur Risikokommunikation, Forschungszentrum Jülich, 2000, Heft. 79.

[26] I. Palmlund, Social drama and risk evaluation, in: *Social Theories of risk* (eds. S. Krimsky and D. Golding), Praeger, Westport, CT, 1992.

[27] F. J. Boster and P. Mongeau, Fear-arousing persuasive message, in *Communication Yearbook* (ed. R. N. Bostrom), Sage, Beverly Hills, CA, 1984, Vol. 8, pp. S330–S375).

[28] A. L. Meijnders, C. J. H. Midden and H. A. M. Wilke, Role of negative emotion in communication about CO_2 risks, *Risk Anal.*, 2001, **21**(5), 955–966.

[29] R. S. Baron, H. Logan, J. Lilly, M. L. Inman and M. Brennan, Negative emotion and message processing, *J. Exp. Soc. Psychol.*, 1994, **30**, 181–201.

[30] R. Hogarth, *Judgment and Choice*, 2nd edn, Wiley, New York, 1987.

[31] H. Jungermann, H. Pfisher and K. Fischer, *Die Psychologie der Entscheidung*, Spektrum, Heidelberg, 1998.

[32] N. D. Weinstein, Unrealistic optimism about future life events, *J. Pers. Soc. Psychol.*, 1980, **39**, 806–820.

[33] D. Kahnemann and A. Tversky, Subjective probability: A judgment of representativeness, *Cogn. Psychol.*, 1972, **3**, 430–454.

[34] H. Jungermann and M. Thüring, Causal knowledge and the expression of uncertainty, in *The Cognitive Psychology of Knowledge* (eds. G. Strube and K. F. Wender), Elsevier, Amsterdam, 1993, pp. 53–73.

[35] G. Gigerenzer, Das Einmaleins der Skepsis, *Über den richtigen Umgang mit Zahlen und Risiken*, Verlag, Berlin, 2002.

[36] P. Slovic, J. Monahan and D. M. MacGregor, Violence risk assessment and risk communication: The effects of using actual cases, providing instructions, and employing probability vs. frequency formats, *Law Hum. Behav.*, 2000, **24**(3), 271–296.

[37] V. F. Reyna, Class inclusion, the conjunction fallacy, and other cognitive illusions, *Dev. Rev.*, 1991, **11**, 317–336.

[38] V. F. Reyna and C. J. Brainerd, Fuzzy-trace theory and framing affects in choice: Gist extraction, truncation, and conversion, *J. Behav. Decis. Making*, 1991, **4**, 249–262.

[39] D. Hattis, Scientific uncertainties and how they affect risk communication, in *Effective Risk communication. The Role and Responsibility of Government and Non-government Organizations* (eds. I. V. T. Covello, D. B. McCallum and M. T. Pavlova), Plenum Press, New York, 1989, pp. 117–126.

[40] D. Kahnemann and A. Tversky, Prospect theory: An analysis of decision under risk, *Econometrica*, 1979, **47**(2), 263–291.

[41] W. A. Magat, W. K. Viscusi and J. Huber, Risk-dollar tradeoffs, risk perceptions, and consumer behavior, in *Learning About Risk* (eds. W. K. Viscusi and W. A. Magat), Harvard University Press, Cambridge, MA, 1987, pp. 83–97.

[42] G. Gigerenzer, The psychology of good judgment: Frequency formats and simple algorithms, *Med. Decis. Making*, 1996, **16**, 273–280.

[43] U. Hoffrage and G. Gigerenzer, Using natural frequencies to improve diagnostic inferences, *Acad. Med.*, 1998, **73**, 538–540.

[44] E. R. Stone, J. F. Yates and A. M. Parker, Risk communication: Absolute versus relative expressions of low-probability risks, *Organ. Behav. Hum. Dec.*, 1994, **60**, 387–408.

[45] D. F. Halpern, S. Blackman and B. Salzman, Using statistical risk information to assess oral contraceptive safety, *Appl. Cogn. Psychol.*, 1989, **3**, 251–260.

[46] W. Brun and K. H. Teigen, Verbal probabilities: Ambiguous, context-dependent, or both? *Organ. Behav. Hum. Dec.*, 1988, **41**(3), 390–404.

[47] D. V. Budescu and T. S. Wallsten, Consistency in interpretation of probabilistic phrases, *Organ. Behav. Hum. Dec.*, 1985, **36**, 391–405.
[48] K. Fischer and H. Jungermann, Zu Risiken und Nebenwirkungen fragen Sie Ihren Arzt oder Apotheker: Kommunikation von Unsicherheit im medizinischen kontext, *Zeitschrift für Gesundheitspsychologie*, 2003, **11**(3), Sonderdruck.
[49] L. M. Moxey and A. J. Sanford, *Communicating Quantities: A Psychological Perspective*, Lawrence Erlbaum Associates, Hillsdale, England, NJ, 1993.
[50] K. H. Teigen and W. Brun, The directionality of verbal probability expressions: Effects on decisions, predictions, and probabilistic reasoning, *Organ. Behav. Hum. Dec.*, 1999, **80**(2), 155–190.
[51] K. H. Teigen and W. Brun, Ambiguous probabilities: When does $p = 0.3$ reflect a possibility, and when does it express a doubt? *J. Behav. Decis. Making*, 2000, **13**, 345–362.
[52] B. B. Johnson and P. Slovic, Presenting uncertainty in health risk assessment: Initial studies of its effects on risk perception and trust, *Risk Anal.*, 1995, **15**(4), 485–494.
[53] K. Fischer and H. Jungermann, Rarely occurring headaches and rarely occurring blindness: Is rarely = rarely? *J. Behav. Decis. Making*, 1996, **9**, 153–172.
[54] D. V. Budescu and T. S. Wallsten, Processing linguistic probabilities: General principles and empirical evidence, *The Psychol. Learn. Motiv.*, 1995, **32**, 275–318.
[55] R. Beyth-Marom, How probable is probable? A numerical translation of verbal probability expressions, *J. Forecasting*, 1982, **1**, 257–269.
[56] D. A. Clark, Verbal uncertainty expressions: A critical review of two decades of research, *Curr. Psychol.: Res. Rev.*, 1990, **9**(3), 203–235.
[57] S. Fillenbaum, T. S. Wallsten, B. L. Cohen and J. A. Cox, Some effects of vocabulary and communication task on the understanding and use of vague probability expressions, *Am. J. Psychol.*, 1991, **104**(1), 35–60.
[58] E. U. Weber and D. J. Hilton, Contextual effects in the interpretations of probability words: Perceived base rate and severity of events, *J. Exp. Psychol.: Hum. Percept. Perform.*, 1990, **16**(4), 781–789.
[59] M. Jablonowski, Communicating risk, words or numbers? *Risk Mange.*, 1994, **41**(12), 47–50.
[60] M. Tavana, D. Kennedy and B. Mohebbi, An applied study using the analytic hierarchy process to translate common verbal phrases to numerical probabilities, *J. Behav. Decis. Making*, 1997, **10**(2), 133–150.
[61] B. B. Johnson and P. Slovic, Lay views on uncertainty in environmental health risk assessment, *J. Risk Res.*, 1998, **1**, 261–279.
[62] D. V. Budescu and T. S. Wallsten, Dyadic decisions with numerical and verbal probabilities, *Organ. Behav. Hum. Dec.*, 1990, **46**, 240–263.
[63] T. S. Wallsten, D. V. Budescu, A. Rapoport, R. Zwick and B. Forsyth, Measuring the vague meanings of probability terms, *J. Exp. Psychol.: General*, 1986, **115**(4), 348–365.
[64] A. C. Zimmer, Verbal vs. numerical processing of subjective probabilities, in *Decision Making under Uncertainty* (eds. R. W. Scholz), North-Holland, Amsterdam, 1983, pp. S159–S182.
[65] I. Erev and B. L. Cohen, Verbal versus numerical probabilities: Efficiency, biases, and the preference paradox, *Organ. Behav. Hum. Dec.*, 1990, **45**, 1–18.
[66] A. Jaffe-Katz, D. V. Budescu and T. S. Wallsten, Timed magnitude comparisons of numerical and nonnumerical expressions of uncertainty, *Mem. Cogn.*, 1989, **17**, 249–264.
[67] K. H. Teigen, To be convincing or to be right: A question of preciseness, in *Lines of Thinking: Reflection on the Psychology of Thought* (eds. K. Gilhooly, M. Keane, R. Logan and G. Erdos), Wiley, Chichester, 1990.
[68] I. Yaniv and D. P. Foster, Graininess of judgment under uncertainty: An accuracy-informativeness tradeoff, *J. Exp. Psychol.: General*, 1995, **124**, 424–432.

[69] T. S. Wallsten, S. Fillenbaum and J. A. Cox, Base-rate effects on the interpretations of probability and frequency expressions, *J. Mem. Lang.*, 1986, **25**, 571–587.
[70] S. McMahan, K. Witte and J. Meyer, The perception of risk messages regarding electromagnetic fields: Extending the extended parallel process model to an unknown risk, *Health Commun.*, 1998, **10**(3), 247–259.
[71] H. Neus, I. Ollroge, S. Schmidt-Höpfner and A. Kappos, Zur Harmonisierung gesundheitsbezogener Umweltstandards – Probleme und Lösungsansätze, in *Umweltbundesamt (Ed.). Aktionsprogramm Umwelt und Gesundheit: Forschungsbericht* (Teil 1), Erich Schmidt, Berlin, 1998.
[72] K. M. Thompson, Variability and uncertainty meet risk management and risk communication, *Risk Anal.*, 2002, **22**(3), 647–654.
[73] B. B. Johnson, Further notes on public response to uncertainty in risks and science, *Risk Anal.*, 2003, **23**(4), 781–789.
[74] P. M. Wiedemann, A. T. Thalmann and M. A. Grutsch, The impact of precautionary measures and scientific uncertainties on laypersons' EMF risk perception, Final Research Report, Research Foundation Mobilecommunication, 2005. Available under http://www.mobile-research.ethz.ch
[75] P. M. Wiedemann and H. Schütz, The precautionary principle and risk perception: Experimental studies in the EMF are, *Environ. Health Perspect.*, 2005, **113**(4), 402–405.
[76] WHO, Harmonization Project. Risk Assessment Terminology, (2004). (in Available under: http://www.who.int/ipcs/publications/methods/harmonization/en/terminol_part-II.pdf (Stand vom 20.06.2005)).
[77] B. Fischhoff, Communicate to others, *Reliab., Eng. & Sys. Safe.*, 1998, **59**, 63–72.
[78] A.T. Thalmann, (2005). *Risiko Elektrosmog. Wie ist Wissen in der Grauzone zu kommunizieren?* Weinheim: Beltz-Verlag.
[79] J. Christensen-Szalanski, and C. Willham, The hindsight bias: A meta-analysis. Organ. Behav. Hum. Dec. Proc., 1991, **48**, 147–168.
[80] B. B. Johnson and P. Slovic, Presenting Uncertainty in Health Risk Assessment: Initial Studies of its effects on Risk Perception and Trust. *Risk Anal.*, 1995, **15**(4), 485–494.
[81] B. B. Johnson and P. Slovic, Lay views on uncertainty in environmental health risk assessment. *J. Risk Res.*, 1998, **1**, 261–279.

European Analysis of the Various Procedures Existing to Interrupt a Clinical Research Protocol Thanks to a French Example of Gene Therapy

Jacques-Aurélien Sergent,[1,5,6] Grégoire Moutel,[1,5] Josué Feingold,[1,3] Hervé de Milleville,[4] Eric Racine,[2] Hubert Doucet[2] and Christian Hervé[1,5]

[1]*Forsensic Medicine and Medical Ethics Lab, 45 Rue des Saints Pères–Medicine Faculty–University René Descartes-Paris 5, France*
[2]*Bioethics Program, 4333 Queen Mary Street, Montréal University QC Canada*
[3]*Honored Director of INSERM Unity 393: "Handicaps génétiques de l'enfant", Hôpital Necker-Enfants Malades Université Denis Diderot – Paris VII France*
[4]*Laboratoire en Sciences de Systèmes d'Information (LASSI), Director, EISTI, Cergy-Pontoise, France*
[5]*International Institut of Research in Ethics and Biomedicine France-Québec, 45 Rue des Saints Pères, Paris, France*
[6]*Groupe de Recherche sur les Physiopathologies Hépatiques Héréditaires (GRP2H), Université de Cergy-Pontoise, France*

Abstract

The decision to interrupt a protocol is deeply dependent on the benefice/risk ratio, especially in the gene therapy field of research where biologists and physicians are directly related. Through the 350 gene therapy researchers selected, because they have presented a research poster during the XIth annual congress of the European Society of Gene Therapy in Edinburgh (November 2003), 62 answered our online anonymous questionnaire. Some important possible axis of regulation criteria and processes were underlined. The online version of this questionnaire is still available on our website: http://sergent.jacques.free.fr.

Questioned about three main research axis (criteria suggested, legitimate and proper process to regulate research and the necessity of a moratorium on somatic gene therapy), the gene therapy community has demonstrated the importance of a proper regulation. Suggesting the Jesse Gelsinger case (1999) and the Bubble kids of Necker (2002), they have also demonstrated their wish to see a regulation process involving a committee, particularly because of their transnational and multidisciplinarity qualities. The possible interactions with not only the patients but also the associations of patients were also pointed out. A regulation mediated by the national ministries was the unique process that encountered more than 50% of approval. The necessity to have a European regulation through a committee has been demonstrated. In fact, about 75% of the researchers encourage a European regulation and approve the extension of a decision taken in one European country to the European community, if this decision is scientifically demonstrated for a multi-country shared protocol. Nevertheless, only 50% of the population approve the creation of a European law, which cannot be adapted to each individual decision. The inter-activity, which exists in a committee, is the characteristic mentioned by the population, determinant in a possible European regulation.

The unawareness of the actual charters and declarations with interest in gene therapy has also been shown by this study, especially by the directors of laboratories or studies.

Keywords : gene therapy, ethics, regulation, trials, interruption, decision.

Contents

1. Introduction 224
2. Material and methods 225
3. Results 225
4. Discussion 227
5. Conclusion 227
References 228

1. INTRODUCTION

Gene therapy became a question in biological medicine since the 1976 [1] Asilomar conference. The first clinical trial in 1990 [2] was about Adenosine Deaminase Deficiency (ADA), which is due to the lack of an enzyme required for the immune system. This monogenic disease was chosen because of its pre-existing background. The first trials on Humans were for Haemophilia B, ADA and an LDL deficiency. In 1999, the Jesse Gelsinger case [3] involving a young man suffering from a non-lethal ornithine transcarbamylase deficiency was the first gene therapy research that was stopped. In fact, Jesse Gelsinger was included in the trial even if he was suffering from a non-lethal form of this disease and the protocol was designed according to his young age [4]. The dose, which was injected to Jesse Gelsinger, was also higher than the upper limit in the protocol designed. Thanks to the lessons learnt from the Gelsinger case, the Gene Therapy research community continues to investigate into various diseases. The "Necker-Enfants Malades" Hospital, in Paris was one of the centres of excellence for research in SCID X1 [5], the Syndrome of Combined Immune Deficiency linked to the X chromosome. The "Fischer Team" worked on the possibility to heal this disease [6] by repairing the immune system and, in 2001, this team declared some positive results for their "Bubble kids". But, in 2002 [7] and more recently in January 2005, the Fischer team had to announce that some of the younger patients had developed a leukaemia due to the localisation of the insertion of the vector utilised. The protocol was analysed by a French ethics committee called CPP (*Comité de Protection des Personnes*). This declaration provoked an interruption in US and Germany, where the same protocol was investigated but in UK, the Gene Therapy Advisory Committee (GTAC) did not recommend the interruption of this particular protocol. Following the French and US interruption a proposition of moratorium was proposed, via an editorial in *Molecular Therapy* in January 2003 [8].

Our study aimed at defining the process of responsibility existing when a decision is required to interrupt a protocol and the eventual internationalisation of this decision. Our research in Medical Ethics was to define the ethical stakes existing during the decision process to interrupt a protocol. Our study was limited to the somatic gene therapy protocol, since only these have been applied to humans.

2. MATERIAL AND METHODS

In order to question researchers, MD and various actors involved in gene therapy research and application, we designed a HTML anonymous questionnaire accessible via a website. This quiz aimed at questioning researchers involved in various phases of research. In order to contact researchers, we decided to question all the researchers who were selected and had presented a research poster during the XIth annual congress of the ESGT, the European Society of Gene Therapy in Edinburgh from the 14th to 17th of November, 2003. In order to prepare this quiz, we realised an assessment of not only the various fears and hopes surrounding gene therapy, but also the general perception of gene therapy and the existing reflection on decisions to interrupt a research protocol. Sixty two research members each from a different research team answered this quiz.

3. RESULTS

The analysis of our data has permitted to underline different results. First of all, it was possible to indicate that the population was mainly in favour of the participation of the research team, the patients, a multidisciplinary committee and persons representing either the ministry or a national agency involved in research protocol regulation (Table 1).

Table 1. Who should be involved in gene therapy trial regulation?

	Total (N=62)	
	N	%
Research team only	35	56.4
Patient	39	62.9
Ministry or National Agency	41	66.1
Promoter	13	20.9
Committee	36	56.4
Patient Association	19	30.6
Other	10	16.1

When the same population was questioned about the actor or association of actors which could legitimately decide to interrupt a research protocol, the two different possibilities which were selected were the ministry or a national agency and the shared decision between the research team and a multidisciplinary committee composed of scientific and non-scientific persons (Table 2).

Finally, when the gene therapy researchers are questioned about the various criteria that could legitimate the interruption of a protocol, 66% approve an interruption existing after a serious and unexpected event such as the death of one of the patients. More than 50% of this population approve an interruption following the emergence of multiple incidents. Interestingly, only one-fourth of the researchers declared being in favour of an interruption due to an alarm caused by an unmatched published result such as the non-action of a molecule, a reaction mainly justified by the various contradictions existing in research and the wish to end their study with their own criteria (population, age, dose ...) (Table 3).

Table 2. Who should decide to interrupt a gene therapy trial?

	Total (N=62)	
	N	%
Research team only	15	24.1
Ministry or National Agency	36	58.1
Promoter	3	4.8
Committee	7	11.2
Research team and patient	30	48.3
Research team and committee	41	66.1
Research team and patient association	16	25.8
Other(s)	4	6.4

Table 3. Which criteria could legitimate the interruption of a gene therapy trial?

	Total (N=62)	
	N	%
Serious and unexpected incident	41	66.1
Less serious incident	33	53.2
Multiples incidents	37	59.7
Alarm caused by an unmatched published result	16	25.8
Other	4	6.4

4. DISCUSSION

This study helped to identify various criteria pointed out by the researchers. So, when an unexpected and serious event happened, about two-thirds of the population declared being in favour of the interruption of the protocol, a serious and expected event is defined by the death of a patient or a highly allergic reaction. When multiple, less serious events occur, about 53% of the population is also in favour of an interruption of a protocol, but the correlation between these incidents and the protocol has to be clearly demonstrated. In the case of a decision, that is to say a responsibility, Hans Jonas [9] declares that each person has "*to consult its fears before its hopes*". That is to say that the benefice/risk ratio has to be evaluated before each decision. A codification of the criteria, helping in the evaluation of the benefice/risk ratio could be developed in the future in gene therapy whose goal is the long-term modification of a gene correcting a non-functional molecule. So, when a physician decides to include a person to a protocol or to interrupt a protocol he needs to keep in mind the possible benefit that other patients could have if the protocol continues. Pre-existing rules could help physicians and researchers when they are facing such a dilemma. Hannah Arendt [10] evoked the *"sensus communis"*, which is a decision taken for the community. About 90% of the population is in favour of the creation of a multidisciplinary independent European and transnational committee involved and specialised in the regulation of gene therapy (data not shown). That is to say that the researchers are developing a reflexion on the decision to interrupt a protocol, temporarily. Their main wish is to develop a European tool, like this committee composed of scientific (physician and biologists), association of patient, patients when applicable, person involved in regulation (like jurists) and some persons from the population.

5. CONCLUSION

In conclusion, we can declare that the research community in gene therapy is developing a common reflection on criteria and regulation on the interruption of protocols. They are also investigating the possibility of a European committee that could legitimately decide when to interrupt a protocol. This harmonisation, obtained by a unique board in Europe responsible for the regulation of a clinical research protocol could also help to determine whether a decision could be internationalised or not. This committee and its precise function could be developed during a workshop whose aims could be to clearly define the criteria, the acceptable risk factors and the various actors who will be involved in the consultation prior to the effective process of decision to interrupt a protocol.

REFERENCES

[1] B. J. Gardner, The potential for genetic engineering: A proposal for international legal control, *VA J. Int. Law*, 1976, **16**(2), 403–29.
[2] W. F. Anderson, Human gene therapy, *Nature*, 1998, **392** (6679 Suppl.), 25–30.
[3] L. Walters, *Why did Jessie Die?* Center for Christian Bioethics, 21 Juillet, 2003.
[4] J.-C. Dumon, M. Sneyers and W. Moens, Rapport décès rapporteé de thérapie génique. Institut Scientifique de la santé publique, 2000.
[5] A. Fischer, S. Hacein-Bay, F. Le Deist, G. De Basile and M. Cavazzana-Calvo, Gene therapy for human severe combined immunodeficiencies, *Isr Med. Assoc. J.*, 2002, **4**(1), 51–54.
[6] S. Hacein-Bay, F. Yates, J. P. Villartay, A. Fischer and M. Cavazzana-Calvo, Gene therapy of severe combined immunodeficiencies: From mice to humans, *Neth. J. Med.*, 2002, **60**(7), 299–301.
[7] Reuters. Des chercheurs francais confirment 2 cas de leucémie liés à la thérapie génique, 05 Mai 2003.
[8] I. Verma, A voluntary moratorium? *Mol. Ther.*, 2003, **7**(2), 141.
[9] H. Jonas, *Le principe responsabilité*, Flammarion, Paris, 1998.
[10] H. Arrendt, Vers une Politique de la Responsabilité, lecture de Hannah Arrendt par Myriam Revault d'Allomes. Esprit 1994, Novembre.

The Future of Public Perception of Gene Therapy in Europe, an Educated Guess

Markus Schmidt

Institute of Risk Research, University of Vienna, Austria

Abstract

Public perception of the opportunities and risks of gene therapy is taking shape in Europe, and various applications of biotechnology are being perceived by the European public. Bio- and gene-technology applications in the medical area are generally seen more positively than applications in the food and agricultural sector (e.g. GM-food). These inclinations, however, cannot sufficiently be explained by different distributions of risks and uncertainties in these areas, instead, the main reason for supporting medical applications may rather be found in the expected future benefits on the individual level. In an attempt to find scenarios regarding the future perception of gene therapy, the following factors may play an important role: the continuous differentiation between applications of gene therapy (e.g., body or germ cells; physical, mental, or behavioural characteristics; and diseases or aesthetics) and their different evaluations; increasing complexity in the post-genomic era including shifting perceptions on (genetic) determinism and perceived control over this new technology; the availability of trustworthy stakeholders and useful information on the issue to make informed decisions; and the possibility that the discourse is being used by some societal interest groups pursuing their particular worldviews. Finally, some brief ideas on how these factors could be dealt with are presented as an input for future discussions on gene therapy.

Keywords: public perception, red and green biotechnology, perceived benefit, differentiation, genetic determinism, worldview, interest groups, trust.

Contents

1. Introduction: risk perception 230
2. Risk perception towards red and green biotechnology 231
3. Possible factors that will shape future (risk) perception 236
   3.1. Continuous differentiation 236
   3.2. Genetic determinism and complexity 239
   3.3. Genetic determinism and reflexive modernisation 240
   3.4. Public discourse 241
   3.5. Trustworthy stakeholders and useful information 242
4. Worldviews, fractal discourse and terra incognita 243
5. Brief ideas on management 245
Annex A. Parts of the danish consensus conference on gene therapy (1995), statements on risk, uncertainty and attitudes towards gene therapy 246
References 247

1. INTRODUCTION: RISK PERCEPTION

Risk is perceived not solely by technical parameters and probabilistic numbers, such as expressed in the formula *Risk* = *Probability* × *Damage*, but also in our psychological, social and cultural context. Individual and social characteristics form our risk perception and influence the way we react towards risks. Our perception is attenuated or amplified in a typical pattern described, e.g., by the psychometric paradigm [1] or by the mental model [2]. These patterns must be taken into account in dealing and working with risk or risky human activities or natural events. It is important to understand the "soft facts" that, especially in case of lay people, outdo the "hard facts" such as technical or medical expertise [3]. Experts typically define risk strictly in terms of annual mortalities or additional costs. Lay people almost always include other factors in their definition of risk, such as catastrophic potential, equity (i.e., whether those receiving benefits from the technology bear their share of risks), effects on future generations, controllability and voluntariness. These differing conceptions often result in lay people assigning relatively little weight to risk assessments conducted by technical experts or government officials [4]. This does not mean that lay people are not aware of, or that they badly estimate, annual fatalities [5,6]. Lay people use other criteria in evaluating risks. Experts include statistical data such as annual fatalities more frequently in their assessment of risks, but they also seem to be prone to many of the same biases as those of the general public, particularly when they are forced to go beyond the limits of available data and rely on intuition and extrapolation [7,8]. See Table 1 for an overview of the most important psychological and social factors influencing risk perception.

Table 1. Psychological aspects attenuating or amplifying the perception of risk.

Attenuate risk perception	Amplify risk perception
Familiar	Exotic
Individual control	Controlled by others
Natural	Manmade
Statistical	Catastrophic
Clear benefits	Little or no benefit
Risks and benefits fairly distributed	Unfairly distributed
Voluntary	Imposed
Information by trusted sources	Information by untrusted sources
In the media	Not in the media

2. RISK PERCEPTION TOWARDS RED AND GREEN BIOTECHNOLOGY

Risk perceptions of a series of biotechnology applications were examined in a public (non-expert) sample and an expert sample [9]. Compared with the experts, the public perceived all biotechnology applications as more risky, a result that can be observed over a wide range of different technologies. An interesting aspect, however, was that both groups perceived food-related applications to be riskier than medical applications. Other studies have previously found similar attitudes towards medical and food/agricultural applications. In the Eurobarometer study [10] carried out in several European countries, respondents asked whether they thought the applications of biotechnology were:

- useful for society,
- risky for society,
- morally acceptable, and
- whether they should be encouraged.

A total of six applications of biotechnology were presented:

- *Genetic testing*: using genetic tests to detect inheritable diseases such as cystic fibrosis mucoviscidosis, thalassaemia.
- *Cloning human cells*: cloning human cells or tissues to replace a patient's diseased cells that are not functioning properly, for example, in Parkinson's disease or forms of diabetes or heart disease.
- *GM-Enzymes*: using genetically modified organisms to produce enzymes as additives to soaps and detergents that are less damaging to the environment.
- *Xenotransplantation*: introducing human genes into animals to produce organs for human transplants, such as into pigs for human heart transplants.
- *GM-Crops*: taking genes from other species and transferring them into crop plants to increase resistance to insect pests and
- *GM-food*: using modern biotechnology in the production of foods, for example to increase the amount of proteins, keep longer or change the taste.

The results showed clearly that applications that would assist fighting diseases and could directly improve our health were more readily accepted than those affecting food production. Although no application for gene therapy was used in the survey, another application namely genetic testing was judged positively in all 15 country surveys, resulting in the strongest support of any of the six applications (see Table 2).

In addition to these kinds of straightforward surveys with closed questions, more in-depth discussion on the issue have been carried out in several

Table 2. Level of support and opposition in European countries for six biotech applications in 2002. (++) Strong support; (+) Weak support; (−) Weak opposition; (−−) Strong opposition

Country	Genetic tests	Clone human cells	Enzymes	Xeno	Crops	Food
Spain	++	++	++	+	++	+
Portugal	++	++	+	+	+	+
Ireland	++	+	+	+	+	+
Belgium	++	+	+	+	+	−
Sweden	++	++	+	+	−	−
Denmark	++	+	+	+	−	−
UK	++	+	+	+	+	−
Finland	++	+	+	−	+	+
Luxembourg	++	++	+	+	−	−−
Germany	+	+	+	+	+	−
Italy	++	++	+	+	−	−
Netherlands	+	+	+	+	+	−
France	++	++	−	+	−	−−
Greece	++	+	+	−	−	−−
Austria	+	+	+	−	−	−

(*Source*: from Ref. [10]).

so-called consensus conferences (CC). These conferences typically involve a small number of lay people that discuss a specific issue with the support of relevant experts (for clarifying questions). Such conferences pretend to gather the "public opinion", which maybe to stimulate political decisions or support management with societal conflicts. In the field of biotechnology most of the CCs carried out so far focused on GM-food or GM-crops, however, at least one CC could be found dealing explicitly with gene therapy (see Table 3).

Representative statements drawn from seven CCs on risks and uncertainties of genetically modified organisms (GMOs) and GM-food can be seen in Table 4. The reaction towards so-called green biotech applications is relatively uniform, taken the different regional and cultural backgrounds of the CCs. Generally, the attitudes are rather sceptical and mostly a "wait and see" approach is preferred, a notion that was reflected in the precautionary principle towards deployment of GM-crops in Europe (1998–2004). Most authors explain this (lay-) reaction towards a new technology by the fact that uncertainties are high and people in industrialised countries have become relatively risk averse, as they expect more to be lost than to be won. In other words, new technologies are rejected on the basis of their uncertainties and risks. An explanation for this phenomenon might be the theory of reflexive modernisation. Reflexive modernisation starts with the notion of a (western)

Table 3. Overview of consensus conferences on some biotechnology applications

Country	Issue	Year	Source[1]
Denmark	Gene therapy	1995	Loka 2003
Norwegian	Food biotechnology	1996	Loka 2003
France	GM crops	1998	Loka 2003
South Korea	GM food	1998	Loka 2003
Australia	GM food	1999	ABC Net 2003
Canada	Food Biotech	1999	University of Calgary 2003
Switzerland	Genetic technology and nutrition food	1999	TA Swiss 2003b
Denmark	GM food	1999	Loka 2003
New Zealand	GM food	1999	Loka 2003

[1]See: ABC Net 2003. http://www.abc.net.au/science/slab/conscont/report.htm#final; Loka 2003. http://www.loka.org/pages/worldpanels.html; TA Swiss 2003. http://www.ta-swiss.ch/www-remain/reports_archive/publications/1999/ta_p_1_99 e.pdf (*Source*: from Ref. [10]).

contemporary change, where "first" (or industrial) modernity makes room for developments leading to "second" (or reflexive) modernity. Under the rule of first modernity, society was based on the belief that everything can, in principle, be mastered by calculation and is thus controllable. The term "reflexive" modernity refers to the erosions of such beliefs and the accompanying intellectual concepts, life styles and policy patterns due to a loss of traditions. This entails uncertainties and different views of nature. Accordingly, the notion that nature as well as society can be steered at will is increasingly vanishing in the western world, leading also to more cautious approach towards new technologies (see e.g., [11–13]).

Given the uncertainties that are comparable in green and red biotech applications, one would expect similar reactions towards these two applications. If the theory of reflexive modernisations may explain the changes going on in western societies, it should predict a rather sceptical and cautious approach also towards red biotech applications. Proof of this theory could be the Danish consensus conference held in 1995 on gene therapy. In this CC the risks and uncertainties were discussed and the attitudes towards gene therapy became visible. Surprisingly the risks and uncertainties – although comparable to green biotech applications – were judged differently.

As an example, when the lay-panel was asked whether it is "*justifiable to implement gene therapy before DNA has been mapped?*", they responded: "*Previously, in combating diseases researchers have researched and applied medicine and methods whose effects and side effects had not been determined in detail before use, and still satisfying results have been achieved. The history*

Table 4. Typical statements from different consensus conferences on green biotech applications

Country, Issue, Year	Representative statements
Norwegian, Food biotechnology, 1996	Too many uncertainty factors are associated with genetic engineering. (The panel) wants more research on effects of eating genetically modified food.
France, GM crops, 1998	We need to be sure there are no higher risks than natural risks before intensifying this type of farming
South Korea, GM food, 1998	We, the Citizen Panel, believe those researchers to be far too optimistic. We believe this because overconfidence in science may have led scientists to be much too optimistic in their abilities to prevent such hazards.
Australia, GM food, 1999	There is currently a lack of understanding in the general community of the risks and benefits involved in introducing GMOs into the food chain, both short- and long-term.
Switzerland, Genetic Technology and Nutrition Food, 1999	According to the opinion of the panel, an estimation of the long-term effects of genetically modified foodstuffs on human health is not possible at the present time. In order to remedy the lack of knowledge about risks, the Citizen Panel recommends that research looking into the area of ecosystem influence be encouraged.
Denmark, GM food, 1999	But experts strongly disagree on the degree of the effect – and whether or not it is hazardous. The disagreement is not only rooted in science, it also stems from ideologic differences.
New Zealand, GM food, 1999	...many of the consequences of this technology are not completely understood. Although scientists seem confident that their procedures are reliable, and the final product safe, they cannot give any guarantee.

of medicine shows that treatment has always been associated with uncertainties. Researchers believe that gene therapy makes it possible to cure a disease whose defective gene is known. It is not necessary to know the entire genome. So far, tests have not disproved this hypothesis."

Here the panel made it clear that it is not necessary to know all possible effects, and that it is fine to use a hypothesis as long as it is not disproved. In sharp contrast, the GM-food issue is judged the opposite as uncertainties are not accepted and industry has to prove that there is no risk involved.

Another statement in the CC was: "*As the treatment would primarily be used on seriously ill people, it is probably an acceptable risk.*"

Also, the CC panel was asked: "*Can gene vectors spread from laboratories to organisms in the surroundings and impact on them ("Turtle Effect")?*", and the response was "*As weak viruses are used, the risk that gene vectors will spread from laboratories and impact on the surroundings is virtually non-existent. They would simply not be able to survive in the natural competition.*" (See more questions in Annex A.)

If we consider – as mentioned before – that comparable risks and uncertainties occur both in red and green biotech applications, the question is: <u>Why is the perception and attitude towards these applications so different?</u>

The reasons for difference is probably not found in the degree of uncertainty and the extent of risks but the benefits involved. The theory of reflexive modernisation departs from the perspective that people in western societies do not want to gain additional benefits or take opportunities but rather want to avoid losses and reject risks. This is probably true in the case of the food sector as can be seen in the "No" – labels used to define quality food products (e.g., no preservatives, no colorants, and no genetical modification; even the organic label rather tells the consumer that it was not produced the conventional way).

In the health sector the notion is different, as long as people suffer from diseases. One quote (from an unknown source) describes this endless challenge: "*Nobody is healthy, he/she just was not checked well enough!*".

Therefore, improvements in health care and medicine are always seen as benefits and opportunities and not as avoided losses. Also the perceived interaction between benefit and risk is of major importance. Although risks and benefits are positively correlated in the real world, they are negatively correlated in people's perceptions. Affect mediates this negative correlation such that if a person feels bad about a technology, they will perceive greater risks and lower benefits. Alternately, if a person feels good about a technology, they will perceive greater benefits and lower risks (see Fig. 1).

Taking this model as a reference, red biotech applications fall under the categories A and B, while green biotech applications fall under the categories C and D (see Fig. 2).

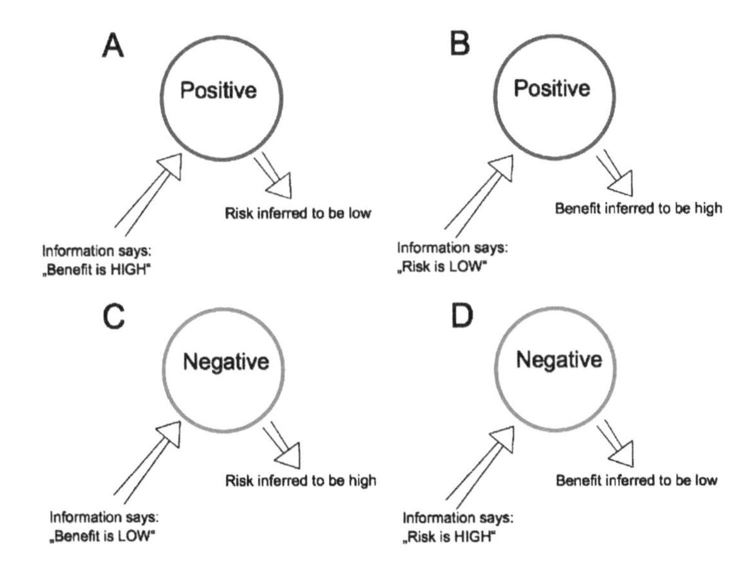

Fig. 1. Model showing how information about benefit (A) or risk (B) could increase the overall affective evaluation in the case of red biotechnology and lead to inferences about risk and benefit that coincide actively with the information given. Similarly information could decrease the overall affective evaluation as in case of green biotechnology (C) and (D) (from Ref. [21]).

3. POSSIBLE FACTORS THAT WILL SHAPE FUTURE (RISK) PERCEPTION

After this brief introduction on contemporary risk perception towards red and green biotech applications, the question remains how risk perception will develop in the future towards medical applications of bio- and gene-technology and especially towards gene therapy. Of course the following chapter can never be more than an educated guess, or to quote the Danish physicist Niels Bohr (1885–1962): "*Prediction is very difficult, especially about the future.*"

Still some ideas on the future shall be discussed in the wake of the "Prospective Technology Assessment" on gene therapy. Some of the factors that come to mind that could be of importance to the future perception of gene therapy are the continuous differentiation between various applications of gene therapy and their particular risk–benefit distribution, the role that genetic determinism and complexity will play in the future, the availability of trustworthy stakeholders and useful information; and the influence of societal stakeholders and movements trying to put forward their particular worldviews.

3.1. Continuous differentiation

The recent Eurobarometer study showed that a majority of Europeans believe that scientific and technological progress will help to cure diseases

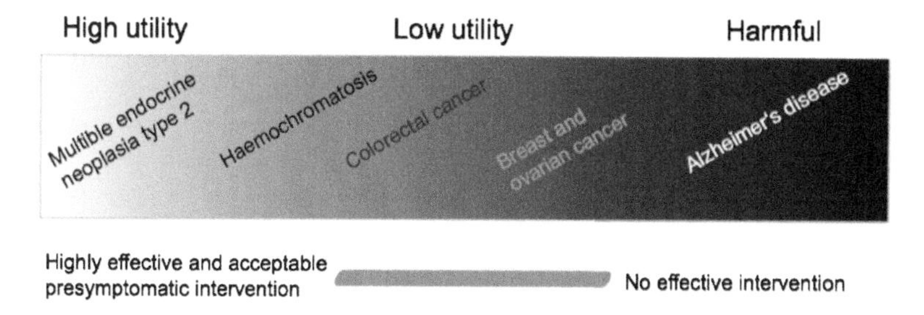

Fig. 2. Utility in predictive genetic testing on five exemplary diseases (from Ref. [14]).

such as HIV-Aids or cancer. Starting from this rather general positive attitude it is likely that attitudes will develop to a more differentiated point of view. Such a differentiation is currently taking place regarding gene testing [14], practically saying that gene testing is only useful in hereditary diseases if a proper treatment is available, otherwise the results are not only not useful but might impose additional psychological stress to the patient (see Fig. 2). Hence the benefits and limitations of a new technology and their socio-economic implications become more and more visible.

On the other hand, over the years more and more reports show an increasingly differentiated picture of the influence of the genome on the (human) phenotype. In a media analysis [15] using US newspaper from 1919 to 1995, it was found that the relative amount of articles describing that *different* genes may have *different* effects on *different* physical and mental state have increased, over articles saying that the cause–effect chains were rather simple (one gene causes one trait) (see Fig. 3). Another form of acknowledging differences – found in the same study – was the different assignment of genetic influence on different human traits (see Fig. 4) such as

- physical characteristics (e.g., size, weight, colour of hair and eyes, some diseases),
- mental characteristics (e.g., intelligence, psychological disorders), and
- behaviour (e.g., personality, criminal, amoral and polygamous behaviour).

Especially interesting is the decrease in beliefs on genetic influence regarding behaviour. (Unfortunately similar studies for Europe could not be found in the literature so far.)

The results shown in Fig. 4 also indicate that gene therapy that interferes with behaviour is unlikely to happen. As long as gene therapy has the aim to cure hereditary diseases the public is likely to accept it, however if mental or behavioural characteristics are at stake the outcome would be rather negative or unclear (imaging the public response to e.g., a genetically engineered piano talent or politician).

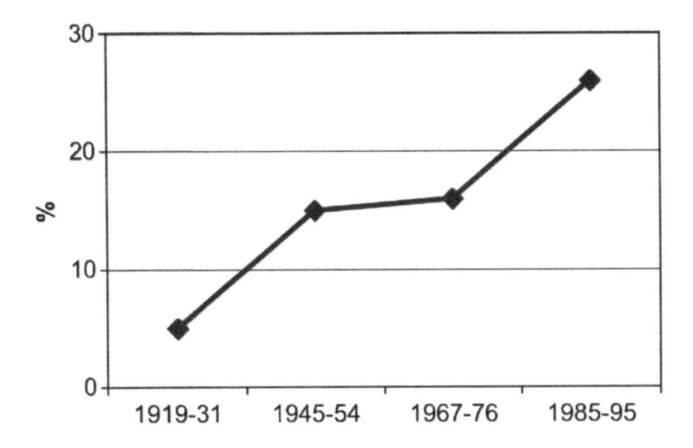

Fig. 3. Share of "differentiated" articles in US Media (from Ref. [15]).

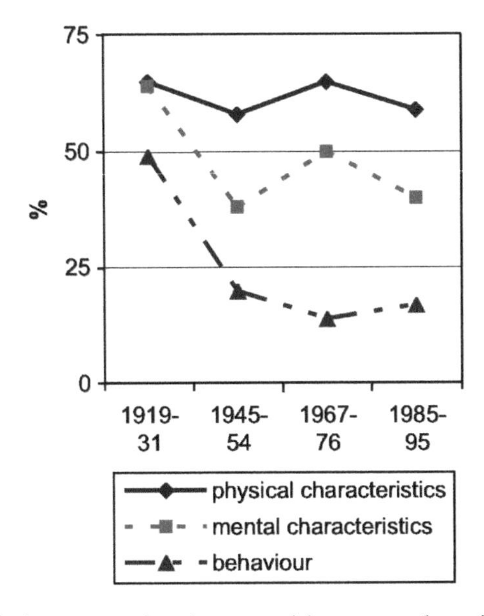

Fig. 4. Characteristics reported to be caused by genes alone in US-Media (from Ref. [15]).

In one case differentiation has already taken place, gene therapy should focus only on somatic cells (body cells) and not on germ cells, and the reproductive parts of the human body. Hereditary of man-made genetic changes in humans[1] will probably remain an ethical taboo.

Differentiation is also an issue regarding the particular risk–benefit distribution of the use of genetic information. As long as genetic information is used

[1] This phrase is a funny pleonasm! In this case "man made" means changes from gene therapy.

with the sole intention to cure diseases (e.g., doctors), people see not so much of a problem. If, on the contrary, personal genetic information is used to pursue interests of third parties (e.g., private industry such as insurance companies) people will very likely refuse such intentions if they have the power to do so.

3.2. Genetic determinism and complexity

As we have seen in the chapter on differentiation the general trend might go (slowly) from a more simple point of view to a more complex one. It is, however, very likely that a notable part of society will not change for the more complex point of view. For reasons that cannot be discussed here the more simple notions of (genetic) fatalism and (genetic) determinism are likely to persist in parts of society as they meet certain pre-existing world-views. As it is foreseeable that this will not affect a minority, this brief chapter is devoted to genetic determinism.

Since the time of ancient Greek philosophers, people asked themselves how our physical appearance and behaviour is shaped. Most of the explanations are oriented on a bipolar nature–nurture model, saying that either all human characteristics are inborn or defined by the genes (nature) or saying that all characteristics are shaped by the environment during lifetime (nurture). This nature–nurture discourse, however, is not restricted to philosophers or scientists but includes practically everybody in our society. It is hardly possible to find anybody who does not have an opinion on this issue.

The successes of molecular biology and genetic research since the discovery of the DNA in 1953 and the recent decoding of the human genome and the beginning of the post-genomic era have triggered the nature–nurture discussion again (see Fig. 5). Relatively new scientific sub-disciplines such as behavioural genetics claim major findings that explain human behaviour by genetics alone. Unsurprisingly, scientific progress aiding genetic determinism also lead to strong criticism, following fears that these

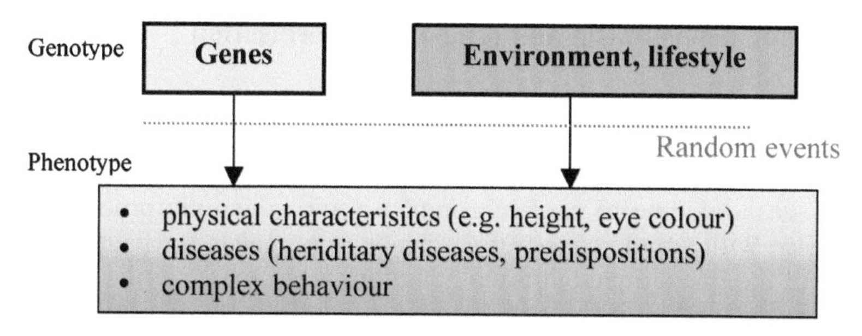

Fig. 5. Nature or nurture, the old question on which of these two "independent" factors shape the phenotype.

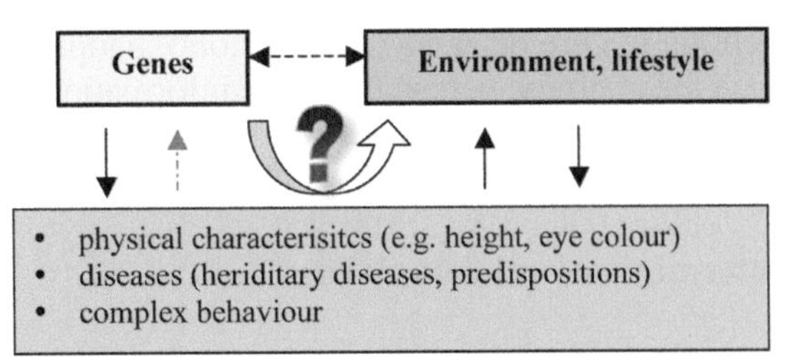

Fig. 6. Nature and nurture interplay in a complex pattern in defining the phenotype. Old notions of determinism are thus not very helpful to explain the formation of the phenotype.

new scientific insights could result in an erosion of human values, free will or moral and legal responsibility ("... my genes made me do it.") [16,22,23].

People defending the "nurture" concept – highlighting the importance of environmental influence – seem to criticise genetic determinism with the aim to uphold these human values. Although the impression is that criticism on genetic determinism is in reality criticism on determinism and fatalistic worldviews as such, as very few comments have been made on environmental determinism.

Beyond nature–nurture [17,25]: To overcome the nature–nurture conflict, one can contemplate what Hebb stated in 1980: "*Our behaviour is defined as 100% by our genes and a 100% by the environment*".

Therefore, the question is not if diseases and behaviour is caused by genes or environment, but rather how genes and environment interact? (see Fig. 6 for an improved concept). By realising the complex interactions among genes, environment and life styles borders are increasingly diluting. Increasing complexities also render it more difficult to define the actual causes of a disease and make precise predictions hardly possible. Especially societal forces promoting the concept of breaking down social borders, e.g., classes, seem to profit from a general non-deterministic worldview, as responsibility as it is understood today is only meaningful in such a context.

3.3. Genetic determinism and reflexive modernisation

Genetic determinism may also be discussed in the context of reflexive modernisation. For example, Allen [16] described the eugenic movement in the USA at the beginning of the 20th century. During this period genetic determinism was welcome to explain complex socio-economic problems (e.g., pauperism) with simple cause-effect chains (e.g., genes for pauperism). At the end of the line simple but brutal solutions (e.g., mass sterilisation) were realised to pretend an increase in perceived control over societal problems. Such an approach falls without doubt in the category of the first or industrial modernisation. For a recent and lucrative attempt to provide simple cause–effect chains in the field of genetic testing [18].

3.4. Public discourse

Although attempts were made to overcome the traditional nature–nurture conflict, the public discourse has only partially responded to increasing complexities. In a study conducted with focus groups in the USA, still 25% of the participants believed in the strongest form of genetic determinism [17]. The problem of genetic determinism is the rejection of active strategies to prevent the risk of diseases (or reduce the probability).

In the media analysis mentioned before [15] it was shown that genetic determinism was fostered by US print media. About 25% of the media reports referred solely to genetic determinism (see Fig. 7). On the other hand, over the years more and more reports showed an increasingly differentiated picture of cause–effect chains.

For clarification it has to be mentioned that only in some rare cases a genetic mutation is certain to cause a disease, whereas in most cases the interplay between genes and environment and lifestyle is of importance. Anyway, even in the post-genomic era, following the encoding of the human genome, it cannot be ruled out that some simple genotype–phenotype interactions will be discovered even for complex behavioural traits.[2]

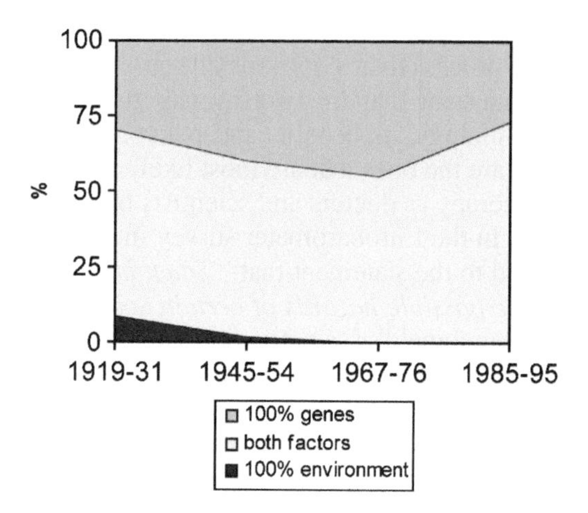

Fig. 7. US Media reports on gene/environment influence.

[2] Example for possible genetic influence on complex behavioural traits "discovery" or "extrovert" gene. The recent "discovery" of an "extrovert" gene is an example of the interplay between quantitative and molecular behaviour genetics. These studies showed that the mean extroversion and novelty seeking scores were higher, while the conscientiousness scores were lower, for individuals with the "long" D4DR (dopamine receptor) allele than for those with the "short" allele. A credible hypothesis of the mechanism of gene action on the phenotypic characteristics is presented, based on the action of the D4D4 gene [24].

3.5. Trustworthy stakeholders and useful information

One of the most important aspects for people living in democracies is the ability to make informed decisions. This basically requires two things: (1) have enough useful information and (2) be able to use this information to make the right decision. In this chapter we do not focus on the decision-making process but discuss the way information is provided and judged.

In risk communication – and management – one of the most important factors is trust. In a complex and highly specialised world people have by no means the chance to be informed on all the issues affecting them. Instead, they rely on the expertise of scientists, regulatory institutions, doctors, non-governmental institutions and other sources of knowledge mostly brought to them by some kind of mass media (TV, radio, print media), the internet and sometimes specific information events. As long as people trust the experts, and those who finally make decisions, and those who watch over the deployment the situation is in most cases calm. Useful and honest information, open and transparent decisions for the benefit of the people are generally well accepted. In case of obvious disinformation, opaque decisions, increasing third party interests and hidden agendas, distrust and opposition is likely to appear. In addition trust has a large asymmetry it takes a long time to be gained, but it can be lost easily [19]. But what role does trust play in informing people? Normally, people tend to accept and use information if it affects them and if the information source is respected and trusted. Information from a trusted source is more likely to influence the decision process than a non-trusted source. For the future of gene therapy it will be important that trustworthy experts tell the public or the patient about this technology, its benefits and risks. In case of gene therapy, doctors and scientists are the ones that are most likely to play this role. This is good news for gene therapy as doctors and scientists belong to the groups that are most trusted [10]. In the Eurobarometer survey in 2002, however, 89% of the respondents agreed to the statement that: "*Scientists ought to keep us better informed about the possible hazards of certain scientific or technological advances*" being a clear signal for scientists to take this role serious.

Another statement found 85.9% agreement of the respondents: "*Scientists ought to communicate their scientific knowledge better*". With *better* they probably mean two things:

- any communication, and
- communication that an interested lay person can follow and understand.

This is basically addressed to scientists and journalists alike, as communicating complex issues in an understandable way is a demanding task. Taken the current situation on the availability of useful information, that gives a good introduction and comprehensive overview, there is still a great demand for structured, comprehensive and useful information. For the economic sector especially in highly dynamic branches "knowledge management" is something they have to deal with. There is no doubt that a lot of information is available (mainly in books, scientific articles and generally spoken on the internet), however, there is very little comprehensive introduction available that is likely to reach the people.

4. WORLDVIEWS, FRACTAL DISCOURSE AND TERRA INCOGNITA

This last chapter is probably the most difficult one. It starts with a simple question: What happens if at the end of the day – even in case most of the other points mentioned so far are solved – we recognised that the whole discussion on gene therapy is not about gene therapy. What, if even in presence of, e.g., a discourse on the risks and benefits of gene therapy, a public discussion on genetic privacy, the appearance of some religious or philosophical moral discourse heavy weight champion telling us what is right and what is wrong, what if all this classical aspects of a neat and nice discussion on gene therapy finally leads us to the conclusion that the discussion is not about gene therapy?

And if I am allowed to explicitly state my own point of view here, I would guess that it is also very likely, that the discussion on gene therapy will hardly be on gene therapy!

But if it is not on gene therapy, what is it then?

After Gill [20] the conflict in the green biotechnologies is not so much on knowledge – as natural scientists would guess – or on risks and interests – as social scientists would guess – but is motivated mainly by different worldviews. The expression "worldview" represents here a set of attitudes towards a number of different issues (e.g., role of humans in nature, political views, personal goals). For German-speaking countries Gill identified mainly three different worldviews[3] that face each other in the discussion:

(1) the conservative, identity oriented,
(2) the utilitarian, progress oriented, and
(3) the romantic, alternative oriented one.

These worldviews come into play if new situations have to be classified and put in context, also if these new situations involve uncertainties that allow several different ways of interpretation. Basically such worldviews[4]

---

[3] Of course people hardly fit entirely into one of these categories, and yes there are also other category attempts (e.g., cultural theory).

[4] A critical mind could argue that putting people into categories or "boxes" does not acknowledge a person's individual character, however the worldview is probably an expression of the individual character.

Table 5. Brief overview of the three worldviews and their perception of nature, diseases, science and technology

Worldview	View of nature
Conservative identity oriented	Nature as the rules of creation, each creature has its place and its identity. Disease as punishment for breaking this rules
Utilitarian progress oriented	Disease as consequence of a hostile nature. Science and technology as means to control nature and find a therapy.
Romantic alternative oriented	Disease because of science and technology, therapy by means of liberation from science and technology.

(*Source*: from Ref. [20]).

give humans a mental tool to structure their complex environment and to help them to – allegedly – understand the world around them, something that creates a feeling of security, rather than remaining in a world with unclear cause–effect chains and no ability to forecast events, thus causing insecurity and fear.

To give an example, Gill cites the different points of view of nature and their relation to science and technology (see Table 5).

In the case of GM-crops the discourse and argumentation frequently followed these worldview conflicts. Interestingly, similar kinds of arguments appeared throughout different levels of the discussion, forming a "fractal-like" pattern of the discourse. As a consequence the observer may draw the conclusion that the discourse on e.g., GM-crops is rather a surrogate conflict of competing worldviews than a conflict of the issue at stake.

A novel issue like gene therapy fulfils at least some of the criteria to become a surrogate conflict between pre-existing worldviews. The novel technology could even be used by some worldview representatives to put forward their demands, claiming the way how to deal with it. To use a metaphor drawn from the world of explorers, the new technology is like a newly discovered island, or continent and each piece of land is going to be conquered, colonised, claimed and defended by the discoverers. The question now is, what kind of post-genomic terra incognita is about to be colonised? Is it mere a *go west* – free for all region, an internationally governed Antarctica type of territory, an arctic icecap, or another worldview surrogate battlefield like in GMO-istan?

A first taste of this *terra incognita* is the recent discussion on gene testing where patient groups and other stakeholders call for a more participative discourse between patients and doctors, instead of the conventional relationship between these two groups. In fact this is not particularly a consequence of the technologies of gene testing, but a wish for reforms in the doctor–patient relationship. A new technology offers the opportunity to install these reforms more easily than to break up old habits and long running "business as usual" (the Collingridge-dilemma). That is why new technologies may be seen also as a barometer for societal consent or dissent, and the way dissent is managed. This is also why a new technology in a sensitive area may be hijacked by interest groups and some stakeholders to pursue their own worldviews and interests.

5. BRIEF IDEAS ON MANAGEMENT

After discussing some of the factors that might play a role in future of public perception (and public discourse) one might ask how to prepare for these. To make it clear, it is not the aim of this article to give management advices. Some very basic ideas, however, may be developed here.

As in most risk communication challenges, also in the case of gene therapy it pays off to follow an open, transparent and proactive strategy in communicating the results and insights of this technology. A slightly improved form of communication, however, is necessary (compared e.g., with GM-crops), one that is able to translate the complex issue of gene therapy into a comprehensive but still "true" content, one that is able to reach to the people, using the communication channels most frequently used by the public, one that is communicated by well respected and trusted individuals, and one that gives the public the opportunity to participate somehow in the way decisions are made. If this is done, a good basis for most of the issues discussed here – regarding knowledge, risks and interests – will be available. In case the issue expands beyond the borders of knowledge and interests, namely to the realm of worldviews, the issue will be much more difficult to resolve. Still, other forms of mediation and conflict managing tools will have to be used.

ANNEX A. PARTS OF THE DANISH CONSENSUS CONFERENCE ON GENE THERAPY (1995), STATEMENTS ON RISK, UNCERTAINTY AND ATTITUDES TOWARDS GENE THERAPY

Question a: Is it justifiable to implement gene therapy before DNA has been mapped?
Answer a: Previously, in combating diseases researchers have researched and applied medicine and methods whose effects and side effects had not been determined in detail before use, and still satisfying results have been achieved. The history of medicine shows that treatment has always been associated with uncertainties. Researchers believe that gene therapy makes it possible to cure a disease whose defective gene is known. It is not necessary to know the entire genome. So far, tests have not disproved this hypothesis.
Question b: Can we be certain that gene therapy will not activate inactive DNA (DNA junk)?
Answer b: To date no problems have been registered in relation to activation of that part of the DNA molecule called DNA junk by a gene implanted through gene therapy. Whether DNA junk can be influenced is not known. Consequently, the risk exists, but experts rate it as small.
Question c: Is there a risk that more than one gene may be injected into the same cell – and which consequences would this have?
Answer c: There is a theoretic risk that more than one gene may be inserted into the same cell, as the gene vector cannot be controlled completely. At worst this could create a mutation which may generate cancer. In the US tests on humans, work is being done on technologies which can trace the implanted gene. A control gene is implanted which allows doctors to intervene and kill the cell.
Question d: Is there a risk that germ cells may be influenced when body cells are treated – and how resistant is our hereditary material to such influence?
Answer d: In general, the treatment of body cells will not affect the germ cells. The risk is limited, but cannot be completely ruled out. If the treatment is effected outside the body, there will be no risk of influencing the germ cells. A similar risk is found in chemotherapy and irradiation. Human genes are very resistant to changes. In a middle-aged person, gene changes may be found in every cell without the person being ill.
Question e: Are there risks of side effects of gene therapy – e.g., that the patient may start to grow again or that viruses (gene vectors) become active?
Answer e: There might be a risk of side-effects in gene therapy, but so far they have not been observed in tests on humans. One side effect of transplantation of several vectors in a cell could be that it metamorphoses into a cancer cell. As the treatment would primarily be used on seriously ill people, it is probably an acceptable risk. However, hypotheses of side effects which, for instance, would cause patients to start growing again, are highly

unlikely. As a weak virus is used as gene vector, the risk that it may become active is minor.

Question f: Can gene vectors spread from laboratories to organisms in the surroundings and impact on them ("Turtle Effect")?

Answer f: As weak viruses are used, the risk that gene vectors will spread from laboratories and impact on the surroundings is virtually non-existent. They would simply not be able to survive in the natural competition.

Question g: Can gene vectors spread from the patient to other persons?

Answer g: Neither of the gene vectors spread from one person to another.

Question h: Will gene therapy on body cells give an increased incidence of hereditary diseases where generation after generation will see more and more persons becoming dependent on gene therapy?

Answer h: In hereditary diseases, where the disease only erupts when the person has had children, gene therapy will not increase the incidence of these hereditary diseases. In hereditary diseases erupting before a person has had children, gene therapy will increase the incidence of these hereditary diseases. This is of course true of all disease treatment.

Question i: Is there a risk of abuse of gene therapy for e.g., genetic warfare and terror actions?

Answer i: As weakened vectors are used, there is no risk of gene therapy abuse for e.g., genetic warfare or terror. In conclusion, it cannot be ruled out that gene therapy may present risks in the long run, as there are very few test results to prove otherwise. Consequently, it will be necessary continuously to evaluate risks to be able to discover any negative side-effects.

REFERENCES

[1] P. Slovic, Perception of Risk, *Science*, 1987 **236**, 280–285.

[2] C. J. Atman , A. Bostrom, B. Fischhoff and M. G. Morgan, Designing risk communications: Completing and correcting mental models of hazardous processes, Part I. *Risk Anal.*, 1994, **14**(5), 779–788.

[3] WBGU, World in Transition. Strategies for global environmental risks. Annual report of the German Advisory council on global change (WBGU), 1998.

[4] V. T. Covello, D. Winterfeldt and P. Slovic, Communicating scientific information about health and environmental risks: Problems and opportunities from a social and behavioral perspective, in *Uncertainty in Risk Assessment, Risk Management, and Decision Making* (eds. V.T. Covello, B. Lester, A. Moghissi and V.R.R. Uppuluri) Plenum Press, New York, 1987.

[5] S. Lichtenstein, P. Slovic, B. Fischhoff, M. Layman and B. Combs, Jugded frequency of lethal events, *J. Exp. Psychol.: Hum. Learn. Mem.*, 1978, **4**, 551–578.

[6] P. Slovic , B. Fischhoff and S. Lichtenstein, in *Rating the risks: The structure of expert and lay perceptions* (eds. V. T. Covello, J. L. Mumpower, P. J. M. Stallen and V. R. R. Uppuluri), Environmental Impact Assessment, Technology Assessment, and Risk Analysis. Band 4, Berlin, Springer, 1985, pp. 131–156.

[7] D. Kahneman, P. Slovic and A. Tversky, *Judgement under Uncertainty: Heuristics and Biases*, Cambridge University Press, Cambridge, 1982.

[8] M. Henrion and B. Fischhoff, Assessing uncertainty in physical constants, *Am. J. of Phy.*, 1986, **54**(9), 791–798.
[9] L. Savadori, S. Savio, E. Nicotra, R. Rumiati, M. Finucane and P. Slovic. Expert and public perception of risk from biotechnology, *Risk Anal.*, 2004, **24**(5), 1289.
[10] Gaskel *et al.,* Eurobarometer 58.0: Europeans and Biotechnology in 2002, 2003, www.cordis.at
[11] A. Giddens. *The Consequences of Modernity*, Stanford University Press, Stanford 1990.
[12] Z. Baumann, *Modernity and Ambivalence*, Blackwell, London, 1992.
[13] U. Beck, *World Risk Society*, Blackwell, London, 1999.
[14] J. P. Evans, C. Skrzynia and W. Burke, The complexities of predictive genetic testing, *Brit. Med. J.*, 2001, **322**, 1052–1056.
[15] C. M. Condit, M. Ofulue and K. M. Sheedy, . Determinism and mass-media portrayals of genetics, *Am. J. Hum. Genet.*, 1998, **62**, 979–998.
[16] G. C. Allen, The social and economic origins of genetic determinism: a case history of the American Eugenics Movement, 1900–1940 and its lessons for today, *Genetica*, 1997, **99**, 77–88.
[17] B. R. Bates, A. Templeton, P. J. Achter, T. M. Harris and C. M. Condit, What Does "A Gene for Heart Disease" mean? *Am. J. Med. Genet*, 2003, **119A**, 156–161.
[18] CPR, Marketing genetics. Consumers' Association in England, 2003, http://www.which.co.uk/
[19] M. Siegrist, Die Bedeutung von Vertrauen bei der Wahrnehmung und Bewertung von Risiken. Akademie für Technikfolgenabschätzung in Baden-Württemberg, Stuttgart, 2001.
[20] B. Gill . Streitfall Natur: Weltbilder in Technik- und Umweltkonflikten, Westdeutscher Verlag, ISBN: 3531138383, 2002.
[21] M. L. Finucane, A. Alhakami, P. Slovic and S. M. Johnson, The affect heuristic in judgments of risks and benefits, *Journal of Behav. Decis. Making*, 2000, **13**, 1–17.
[22] J. S. Alper, Genes, free will and criminal responsibility, *Soc. Sci. Med.*, 1998, **46** (**12**), 1599–1611.
[23] J. S. Alper and J. Beckwith, Genetic fatalism and social policy: the implications of behavior genetics research, *Yale J. Bio. Med.*, 1993, **66**, 511–524.
[24] R. C. Bailey, Hereditarian scientific fallacies, *Genetica*, 1997, **99**, 125–133.
[25] R. Steven, Moving on from old dichotomies: Beyond nature-nurture towards a lifeline perspective. *Br. J. Psychiatry*, 2001, **178** (suppl. 40), pp. 3–7.

Subject Index

A priori probability, 212
AAV, 17–19, 21, 23
Aav serotypes, 17, 19, 21
Abortion, 110, 113, 115
Act for Protection of Embryos, 118
Adeno-associated virus, 17–18
Adenovirus, 33–37, 41–44
Adverse effects, 204, 206
Ambiguity, 206, 211, 213, 215–216
Artificial
  chromosome, 113
  virus-like particles, 47–48
Assisted reproduction, 110, 113
Availability of risk information, 210

Base rate, 212, 214
Beliefs, 209, 216
Bioethics, 143–144, 150, 152, 154, 156–157
Biological barriers, 48
Biomimetic, 61, 70
Brain, 17–18, 20–21, 23
BRCA1 and BRCA2, 159, 164, 167, 169

Canavan's disease, 21
Cancer, 112
Capsid, 20
Charge, 47, 51–52
Clarity, 203, 205, 211, 213, 215, 217
Clinical
  application, 47–48
  trial, 111, 117
Combinatorial, 72
Communication mode preference paradox, 213
Comprehensibility, 203, 207, 211–212
Concept of risk / risk concept, 203, 207–208, 216
Confucianism, 143, 152–153
Conjunction fallacy, 210
Consent, 143, 148, 155
Constraint, 115
Context, 209, 214
Contraception, 114, 118
Controversy, 181, 198–199
Counselling, 109, 114–115, 117
Culture, 143–144, 150–151, 153, 156
Cytotoxicity, 65, 67

Decision, 223–227
Decision making, 203–205, 216–217
Differentiation, 229, 236–239
Disease associated genes, 159
DNA, 57–68, 70–71, 109, 112–113, 119
DNA methylation, 112

Embryo, 118–120, 143, 150–153, 155
Emotionality, 203, 208–209, 217
Endocytosis, 59, 61–62, 65, 69
Enhancement, 111, 117, 123, 128, 132–135, 137–141
Epigenetic, 109, 112–113
Ethical consideration, 89
Ethics, 99, 106–107,143, 148, 153–154, 156, 223–225
Eugenics, 123–128, 131, 133–134, 140–141
Evidence characterization, 203, 214–216
Exogenous DNA, 47–48

Fear appeal, 209
Federal Drug Administration, 116
Fetal gene therapy, 89–94
FMRI, 77–83, 85
Freedom of research, 181

Frequency, 210–214
Functional element, 47, 52
Fundamental right, 109, 111

Gelsinger case, 204
Gene
  delivery, 17, 23
  expression, 112
  therapy, 3, 11–12, 33, 36, 99–107,109–119, 123–124, 127–128, 223–227
Gene transfer mechanism, 47–50, 54
Genetic
  determinism, 229, 239–241
  diseases, 3, 12
  testing, 159–160, 163, 167–170, 175–178
Germ-cell, 99–103
Germline, 109–113, 115–119
Germline gene therapy, 112

Haemophilia, 117
Hazard, 205–206, 208, 214
Heart, 33–44
Heparan sulfate proteoglycan (HSPG), 51
Hereditary diseases, 110
Heteroplex, 47–52, 54
Heuristics, 203, 209–210, 217
Hindsight bias, 210
Human
  dignity, 117–118
  genetics, 114
  rights, 117

Illusion of control, 210
Imaging, 77–86
In vivo, 33–35, 37, 39, 41–43
Incidence, 212
Information processing, 209
Informed, 143, 148, 150, 155–156
Informed consent, 109, 115, 119
Informed decision, 204
Insertional mutagenesis, 112
Interest(s), 143–144, 151, 155–156
Interest groups, 229, 245
International, 143–145, 148, 150–151, 153–155
Interruption, 224, 226–227
Intracellular transport, 50, 53
Intrauterine application, 93
Inventive step, 171
Inverted terminal repeats, 18

Lay people, 181, 187
Layperson, 203, 207, 211, 213, 215, 217
Linkage analysis, 3, 5–6

Media, 204, 208, 211, 217
Medical profession, 89, 91–93, 95
Medicine, 109, 114, 120
Mental model, 210
Misinterpretation of coincidence, 210
Molecular, 77–79, 84–85
Moral, 143–144, 151–153
Morality, 143, 155
Mucopolysaccharidosis, 21
Multidimensionality of risk, 203, 208
Multidisciplinary, 71–72
Mutation, 112, 117

Natural frequency, 211–212, 216
Naturalistic fallacy, 116
Nervous system, 17–21
Neurological disorders, 17–18, 20, 23
Non-viral gene delivery, 57, 62–65, 67–69, 71–72
Nonviral
  gene therapy, 47
  gene transfer, 47–48, 50
Novelty, 159–160, 171–173
Nucleus, 47, 50–51, 53–54
Numerical description of probability, 213

Oncogenesis, 113
Overconfidence, 210

Pain, 77–78, 80–85
Parkinson, 21–22
Particle, 48–49, 51
Patent, 159–162, 165–167, 169–176, 178–179
Patent and genetherapy, 159
Patents on BRCA1 and BRCA2, 162
PEGylation, 66–67
Peptides, 47, 49–52
Perceived benefit, 229
Physician, 89, 91, 93–94
Plasmid, 113
Policy-style, 181–182, 187, 191, 196
Poly(ethylene oxide) (PEO), 65
Poly(L-lysine), 63
Polycations, 57, 62–64, 66–71
Polyethyleneimine (PEI), 64
Polyplex, 62–63, 65–66
Popularisation, 181–183
Preimplantation genetic
  diagnosis, 118
  modification, 118
Prenatal diagnostics, 113
Priority, 159–160, 171–173
Privacy, 77, 84
Probability, 205, 208–214, 216
Prospective political stakes, 181
Public
  opinion, 109, 114
  perception, 229
  problem, 181, 191

Red and green biotechnology, 229
Regulation, 223–225, 227
Relative risk, 205, 211–212
Representativeness, 209
Reproduction, 110, 113, 115–116
Reproductive
  autonomy, 109–110, 113–119
  technologies, 123–124, 126, 128, 134, 139
Research ethics, 99, 106
Responsibility, 110–111, 115–116, 119
Retinal diseases, 3–4
Retinitis pigmentosa, 3, 5, 12
Risk
  assessment, 109, 111, 116–117, 204–208, 214–215, 217
  communication, 203–208, 211, 213–217
  management, 207
  perception, 203–205, 208–209, 212, 215, 217
  stories, 208–209

Science regulation, 181
Self-assembly, 61, 66, 68
Serotype, 17, 19, 23
Side effects, 109–113, 116, 119
Small probability, 211
Social, 143, 152–153, 156
Social amplification, 208
Social coercion, 109–110
Societal context, 181–182
Somatic-cell, 99–101, 103–104
Somatic gene therapy, 109–111, 116, 118
Specificity, 47, 50, 52
Spinal cord, 17, 20–21
Standardization, 47, 51–52
Standards, 143–144, 151, 153–156
Stem cells, 143, 151
Sterilization, 109, 115
Strength of evidence, 203, 211, 214–215, 217
Synthetic polymers, 57–58

Target gene delivery, 90
Technical-scientific risk assessment, 208
Transcriptional regulation, 17, 22
Transfection, 58–59, 61, 64–65, 67–70, 72
Transfection mechanism, 47, 49
Transfection system, 47, 50, 52–54
Transparency, 203, 207, 211, 213, 217
Trials, 224
Tropism, 17, 20

Trust, 229, 242
Tumor, 49

Uncertainty, 203–207, 211–216
Unclear risk, 206
Unrealistic optimism, 209

Vague uncertainty, 203, 211, 215
Vagueness, 206, 211, 213
Variability, 206, 213, 215
Vector, 47–48, 50, 111–112, 117–118
Verbal description of probability, 211, 213
Viral vectors, 48, 58, 61, 65, 69–70
Virtue ethics, 123, 128–133
Virtues, 123, 129, 131, 133–135, 137–140
Virus, 47–48, 53

Worldview, 229, 240, 243–244

Printed and bound by CPI Group (UK) Ltd, Croydon, CR0 4YY
08/05/2025
01865009-0001